GREAT DISCOVERIES IN SCIENCE

The Copernican System

Erik Richardson

New York

Published in 2018 by Cavendish Square Publishing, LLC
243 5th Avenue, Suite 136, New York, NY 10016

First Edition

Library of Congress Cataloging-in-Publication Data

Names: Richardson, Erik.
Title: The Copernican system / Erik Richardson.
Description: New York : Cavendish Square, 2018. | Series: Great discoveries in science | Includes index.
Identifiers: ISBN 9781502627766 (library bound) | ISBN 9781502627773 (ebook)
Subjects: LCSH: Astronomy--Juvenile literature. | Copernicus, Nicolaus, 1473-1543--Juvenile literature.
Classification: LCC QB46.R534 2018 | DDC 520--dc23

Editorial Director: David McNamara
Editor: Caitlyn Miller
Copy Editor: Michele Suchomel-Casey
Associate Art Director: Amy Greenan
Designer: Lindsey Auten
Production Coordinator: Karol Szymczuk
Photo Research: J8 Media

The photographs in this book are used by permission and through the courtesy of: Cover, Andreas Cellarius/File: Cellarius Harmonia Macrocosmica - Planisphaerium Copernicanum.jpg/Wikimedia Commons; pp. 4–5 Camille Flammarion, L'Atmosphère: Météorologie Populaire (Paris, 1888), pp. 163/File: Flammarion.jpg/Wikimedia Commons; pp. 8–9 Ullstein Bild/Getty Images; p. 14 Ken Welch/ Perspectives/Getty Images; p. 18 A. Dagli Orti/DeAgostini/Getty Images; pp. 20, 23, 35 Oxford Science Archive/Print Collector/Getty Images; p. 25 Science & Society Picture Library/Getty Images; pp. 28, 96 Universal History Archive/Getty Images; p. 30 Paul Almasy/ Corbis/VCG/Getty Images; p. 36 Walters Art Museum/File: Fra Carnevale - The Ideal City - Google Art Project.jpg/Wikimedia Commons; p. 39 Workshop of Lucas Cranach the Elder (German, Kronach 1472–1553 Weimar)/Gift of Robert Lehman, 1955/The Metropolitan Museum of Art/Public Domain; p. 47 NYPL/Science Source/Getty Images; p. 48 Traveler1116/E+/Getty Images; p. 50 © Marie Lan-Nguyen/Jastrow (Own work)/File: Aristarchus Diebolt Merley cour Carree Louvre.jpg/Wikimedia Commons; p. 55 Photo12/UIG/ Getty Images; p. 58 Hel-hama (Own work)/File: Woolsthorpe Manor.jpg/Wikimedia Commons; p. 60 De Agostini/G. Nimatallah/Getty Images; p. 63 © Roger Ressmeyer/Corbis/VCG/Getty Images; p. 66 Imagno/Getty Images; p. 68 Leemage/Corbis/Getty Images; p. 74 Das Wortgewand/File: Jupiter and the four Galilean moons (artistic).jpg/Wikimedia Commons; p. 80 English School/Bonhams/File: Isaac Newton, English School, 1715-20.jpg/Wikimedia Commons; p. 83 www.rijksmuseum.nl/File: Jan Verkolje - Antonie van Leeuwenhoek. jpg /Wikimedia Commons; p. 87 Ann Ronan Pictures/Print Collector/Getty Images; p. 88 Mikkel Juul Jensen/Science Source; p. 92 Jastrow (2006)/File: Urban VIII Bernini Musei Capitolini.jpg/Wikimedia Commons; p. 95 File: Albert Einstein 1921 by Ferdinand Schmutzer.jpg /Wikimedia Commons; p. 101 NASA and the European Space Agency/File: Hubble ultra deep field.jpg/Wikimedia Commons; p. 106 ESO/Igor Chekalin/File: Messier 78.jpg/Wikimedia Commons; p. 109 Royal Astronomical Society/Science Source.

Printed in the United States of America

This print is from a famous woodcut showing Giordano Bruno's extraordinary dream about the universe.

Introduction: Where Does Earth Fit in the Universe?

We need to think of the **Copernican Revolution** much like we would a great symphony or a beautiful work of art. Instead of communicating through the forms of music or through light, shadow, and colors, the great contributors use the language of mathematics and **physics**. It is important to keep in mind that talking about the time signature of a symphony, or the way the themes create tension with each other, helps us understand it in one way, but we can miss seeing where it takes us. The same is true of a work of art. You can think about the way the artist uses a certain brush technique to create the effect of light reflecting off the water, but you also have to set that aside at some point and notice what the artist is trying to express. The best artists, the best musicians, are trying to stretch the paints or the notes to capture something new, to carry our minds someplace we haven't been before.

The creative thinkers of the Copernican Revolution were trying to capture and express something, too. To appreciate that, you need to meet a great thinker whose imagination leapt out ahead of where science could reach yet and sparked some ideas that made the other characters in our story want to find a way to stretch science so it could express that new vision.

Giordano Bruno was born in 1548 in a little town near Naples, Italy, and he died in 1600 in Rome. A statue now stands where the Catholic Church burned him at the stake for spreading ideas it did not agree with. When he was thirty years old, this Dominican monk had a dream. In the dream, he found himself in a world that was surrounded by a confining bowl of stars. He was filled with fear, but he went over to the edge and found that the "bowl" was just a curtain. When he lifted the curtain and crawled under it, he was in an expanded universe stretching in all directions. He took flight, with no up or down and no center, and he was overcome to see that the sun was just another star and the other stars were suns and they could each have Earths like ours.

Though deeply troubled by Bruno's idea of an **infinite** universe, German **astronomer** Johannes Kepler, in writing to Galileo, pointed out that the stars, when seen through a telescope, still sparkle, which planets do not. Because of this he asked Galileo what other conclusion they could draw except to agree with Bruno's idea that the stars were suns. When Bruno lifted up that curtain, in his dream, and found a bigger universe outside, he lifted it up for the other thinkers as well.

Bruno did not have the scientific tools to prove his idea, but he sparked the curiosity and the ambition of some thinkers who could create the tools they would need. So, let us turn to meet these great scientists, these musicians of angle and number, inventors of new math, builders of new telescopes, and see how they shifted our understanding of Earth's place in the universe, taking us from **geocentrism** to **heliocentrism**: the Copernican Revolution.

We will start by getting an understanding of the old model of the solar system that the ancient Greeks had passed down. Then we will go on to meet the great thinkers at the center of the revolution. We will learn a little about their lives and their backgrounds, so we will have some understanding of how their paths led them to key turning points and how their paths intersected with each other.

That will give us the context for each of the different steps forward that helped lead everyone from the ancient way of thinking about a very small universe to the modern conception of an infinite universe. We will see how this required new instruments, like the telescope, and how it required a new way of thinking about even basic ideas, like what happens when you drop something and it falls to the ground. Of course, we will also see how this put our scientists at odds with the church as surely as Bruno's dream.

Lastly, we will follow through to look at how that new way of thinking about science and the universe carried forward to shape current research and discoveries as we continue trying to stretch our science further and further.

Here we see how Ptolemy thought the universe was laid out, with the solar system making up most of it.

CHAPTER 1

The Problem of Understanding the Universe

Before we are in a position to fully appreciate how remarkable the Copernican Revolution was, we need to spend some time exploring the history behind it. If we can make sense out of the problem that famous theorists like Nicolaus Copernicus and Kepler were trying to solve, we can understand their goals a little better. Similarly, if we can look at the way earlier thinkers tried (and failed) to solve those same problems, only then can we measure their brilliant solutions by comparison.

Even as we set out to do that it might be hard to connect with the earlier thinkers. We now know that their project to prove that Earth is the center of the solar system was doomed to fail. We also know that their work gets superseded by the ideas of the geniuses who followed. If we look back at them through the lens of their failures, we are doing them and ourselves a disservice. Theorists like Aristotle and Ptolemy were bright and passionate thinkers, their ambitions were no less than Galileo's, and their models were—for their time period—far more ingenious than you or I would likely have come up with.

The RISE and FALL of the PTOLEMAIC MODEL

When the Ptolemaic model was established, in the last couple centuries before the rise of Christianity, it was cutting edge. This model said that Earth sat, unmoving, at the center of the solar system. The Ptolemaic model did an admirable job of predicting the location of various stars and planets as they traveled along their various paths. It was also far better at this than any of the other ancient models. In fact, the Ptolemaic model is still often used in engineering because it's still approximately true.

However, while it might be close enough for some uses, research and observation over the centuries gradually showed more and more places where it was too far off to be good science. In details like the position of planets and the timing of equinoxes, Ptolemy's model was off target. Trying to solve those gaps between the model and observation determined the main focus of many of the astronomers who came after him. They still had confidence, though, that the adjustments to the theory would turn out to work and the model would survive in a new and improved version.

As time went on, the number of places where astronomers had to make adjustments and corrections to get the Ptolemaic model to fit their observations continued to grow. They also came to realize that making an adjustment to one part of the model resulted in creating a new gap in a different part of the model. These gaps meant that what the model predicted and what scientists observed were different. But, because there was not as much communication (or as fast) between scientists back then, it took a long time for the scientific community to realize just how many of these gaps there really were when you put them all together in one list.

By the time the sixteenth century came to pass, more and more of Europe's best astronomers were starting to grasp how often the old model was failing. The standard of good science and good scientific models is to find the rules and patterns that represent parts of the world around us. Great thinkers had started to doubt that anything as messy and full of exceptions to its own rules as the Ptolemaic model could possibly be true of nature. This same perception is what led Copernicus himself to write in the preface to his famous work, *De revolutionibus*, that the traditional model of astronomy had become a kind of monster.

It is valuable to take a minute to understand something important about scientific progress. It seems quite easy for us to look back from our current position and ask why Aristarchus's model of astronomy was not chosen over Ptolemy's. Aristarchus (310–230 BCE) had the insight that the sun, rather than Earth, should be put at the center of the model. However, when Aristarchus's version was first proposed, it was not clear that it had any kind of competitive advantage over the competing model that kept Earth at the center. To put it another way, the geocentric model seemed to work perfectly fine and didn't need to be replaced. It was only after centuries of further astronomical observation and discovery that the geocentric model accumulated so many holes and patches that it needed to be replaced.

Keep in mind that there are always "holes" in our scientific model for understanding things. The whole project is a kind of puzzle, and it is perfectly normal as we work our way along to find missing pieces. It is like finding a bunch of puzzle pieces on a table. We're not sure which box they came out of, so we look through empty boxes on the shelf and form a guess about which box has a picture on the front that seems to match the pieces we're working with. If we have a lot of success with that box front, we feel good about our choice and we carry on, expecting that the picture of the puzzle we have in mind will allow us to eventually fit more of those missing pieces into place. It is only after piling up a certain number of pieces that we can't seem to make fit with the picture on the front of the box that we have been imagining that we realize maybe we are looking at the wrong box.

By the time Copernicus came along, he found that there were many, many mismatched parts of the Ptolemaic picture being developed by astronomers. Copernicus said it was as if an artist brought together images of hands, heads, and feet from different people. While each might have been well drawn, since they were not from the same person, the result of putting them all together looked more like a monster than a normal human.

The WAY SCIENCE WAS DONE BEFORE the REVOLUTION

Before Copernicus, Aristotle can be considered the last significant **cosmologist** and Ptolemy as the last great astronomer. We can talk about some of the small steps forward that were made in the five centuries between Aristotle and Ptolemy, or in the thirteen centuries between Ptolemy and Copernicus, but these were exactly that—small steps. The fact is that those two—Aristotle and Ptolemy—were the dominant authorities of almost all astronomical reasoning and research. In fact, they were still the backbone of university coursework in the time of Copernicus, Galileo, and Kepler.

At the beginning of the sixteenth century, people were working from these models, in part because they had, indeed, been brilliant models and added much to the ability of thinkers to explain things about the world around us. At the same time, part of the reason so little real progress had been made in the intervening centuries had to do with the fact that much of the ancient tradition had been lost for a long time. By the second century BCE, the Mediterranean civilization that had given rise to thinkers like Aristotle had come under the controlling influence of Rome. By the second century CE, figures like Ptolemy and Galen (famous for medical and biological research) had become the last figures doing much more than writing commentaries and encyclopedias on other scientific thinkers.

Muslim and Catholic Influence

At the point in the seventh century when Muslims invaded the region, Europe had slid into the Dark Ages with little meaningful research or progress happening. These invasions caused even further decline, as now Europeans lost contact with even the books and documents of thinkers like Aristotle and Ptolemy. What little was still known of their work (and others in a variety of fields) was in fragments and second-hand discussions in some of the encyclopedias and such. This virtual disappearance of science was amplified by the dominance in the West of the Catholic Church. Throughout the Middle Ages the church pushed science to the margins in favor of religious explanations for the creation of the universe and the world around us. The church's influence had clear consequences for scientists. In fact, Galileo faced house arrest for his scientific theories because they conflicted with church teachings.

As Europe moved out of the Middle Ages in the tenth century and on toward the Renaissance, which was underway by the twelfth century, connections and trade with Islam allowed texts and translations to begin flowing back into Europe. Greek and Hellenistic authors moved once again to the place of honor in the fields of research and education. In a

This map shows how much of the known world was under Muslim control during the period we mistakenly think of as the Dark Ages.

very real sense, then, when Copernicus studied the science and astronomy of Aristotle and Ptolemy in his university classes at the end of the fifteenth century, it was as if he was just starting where they left off.

There was a certain additional problem created as a result of this weird time-lapse effect. To the medieval thinkers, Ptolemy and Aristotle seemed like they were working and researching at the same time, instead of five centuries apart. Therefore, scholars in the Middle Ages and Renaissance tended to overlook certain important differences between the different ways of doing science—astronomy in particular. In a way that would be important to Copernicus, the Ptolemaic approach was more about mathematics and geometry in contrast to Aristotle's focus on certain kinds of philosophical reasoning (like that the reason planets travel in a circle is because they are eternal and circles are a more eternal shape, so that would be the only possible combination).

Science Under Control of the Catholic Church

While science did play a role in the thinking of the late Middle Ages, it was seen as a servant to the interests of theology (religious theory). Not only was the Catholic Church the dominant force in social life and politics, but most of the scholars and university professors were Catholic as well. In fact, the universities themselves were Catholic owned and managed.

We must be careful, though, not to imagine an overly simple description of the church's position. Science was well-taught in the universities, and we must keep in mind that Copernicus himself had an uncle who was a bishop. At the same time, the church's view of what kind of science could be studied, and how research could proceed, was too rigid to allow for much breakthrough work. The Catholic Church was, at first, reluctant to accept ideas of Aristotle, with bans against his scientific work in 1210 and 1215. Thinkers like Albert the Great and Thomas Aquinas, however, showed how Aristotle's ideas could be squeezed and stretched in different ways to fit with Christian doctrine. Then Aristotle became one of the church favorites. The example of Aristotle stands in interesting contrast, however, to other members of the scientific community. In 1616, for instance, the church banned any books that promoted the idea of Earth moving instead of being "fixed in its foundation" as described in the Bible. To make the matter somewhat more complex, we also find Catholic thinkers attempting to intermingle theology and science in novel ways—like accounting for certain motions of the planets by claiming it was the work of angels.

WHAT WERE the OLD MODELS?

With that larger overview in mind, let us now turn to consider the models of Aristotle and of Ptolemy in a little more detail. There were, of course, a great variety of other models competing against these from earliest times. If we were digging deeper with just a careful focus on their periods of antiquity, we would find that there were rivalries and conflict about Earth's position in the universe that would rival those of the Copernican Revolution. However, we will focus on the theories of Aristotle and Ptolemy because each of them was the most successful in his day. That is, they were the ones who came out of the heated war of ideas on top. Their models governed astronomy for the many centuries leading up to Copernicus.

In an interesting way both Aristotle's and Ptolemy's theories start from a common idea. Both of these thinkers would draw the paths of the planets by showing a set of concentric circles with Earth at the center. It was also central to their models that the planets were traveling with something called "uniform angular motion." This means that each planet was thought to travel the same number of degrees around its own circle of movement in a given amount of time. To simplify, we can say uniform angular motion means planets travel at just one speed—they do not speed up or slow down at different places in their orbit. (Today we know this idea is false.) From there, the two thinkers take different paths, though.

Aristotle's Theory

Aristotle lived from 383 to 322 BCE in ancient Greece. In Aristotle's version of cosmology, the universe was made up of a set of large spheres. These spheres all had their center point at Earth, and they were nested. Imagine a set of glass globes, a small one inside a medium one, and that, in turn, inside a larger one (and so on). Then imagine that each of these glass spheres has a small bead embedded into its surface. For Aristotle, the beads are the planets, each lodged in the surface of a sphere, and all of them are spinning around.

There were many problems with this model. Often its predictions were wrong according to astronomers' observations. Also, the Aristotelian model had no ability to explain why a given planet was sometimes brighter than other times since it said that planets were always the same distance from the center. (That has to be true, based on the definition of spheres.) As we will appreciate later, this model also had no

Few thinkers before Newton would have an impact on so many fields as the great philosopher and scientist Aristotle.

way to explain the fact that planets seemed to slow down at times and even change direction as they moved.

Some of the problems of later observations were not considered significant to Aristotle as they would be to Copernicus or to us. Aristotle's idea of science was largely one that could be conducted by careful reasoning while seated in a cozy chair by the fireplace. Now this is a little bit of a simplification, but it serves to help get the point across. His use of reason in and of itself to build explanations of the natural world was inherited from a long line of prominent thinkers before him. In a sense, he was the **paradigm** case to illustrate the ancient Greek tradition of figuring out the nature of the universe.

As an example of his approach to science, he simply took it for granted that planets must experience **friction** because he could not allow for the idea that space was actually empty (a fascinating theory in the history of science that probably deserves its own whole book). Since planets would experience friction, he reasoned, they would have to slow down unless something or some force was continually pushing on them. Partly because of this, he set force equal to speed. This was a far cry from Isaac Newton's later realization that the force of an object had to do with its **mass** and how fast it was speeding up, not just how fast it was traveling.

In a similar fashion, Aristotle's collection of ideas about the universe included the thought that **celestial** objects—suns, planets, stars, and more—had to move in circles because they were spheres and circular movement is the natural form of movement that spheres take. He felt a need for no further explanation than to base his claim on the accepted definition of spheres. In Aristotle's mind, the obviously noncircular path of things like **comets**, then, meant they could not be objects in the heavens. They must be some weird phenomenon happening in the atmosphere, he said.

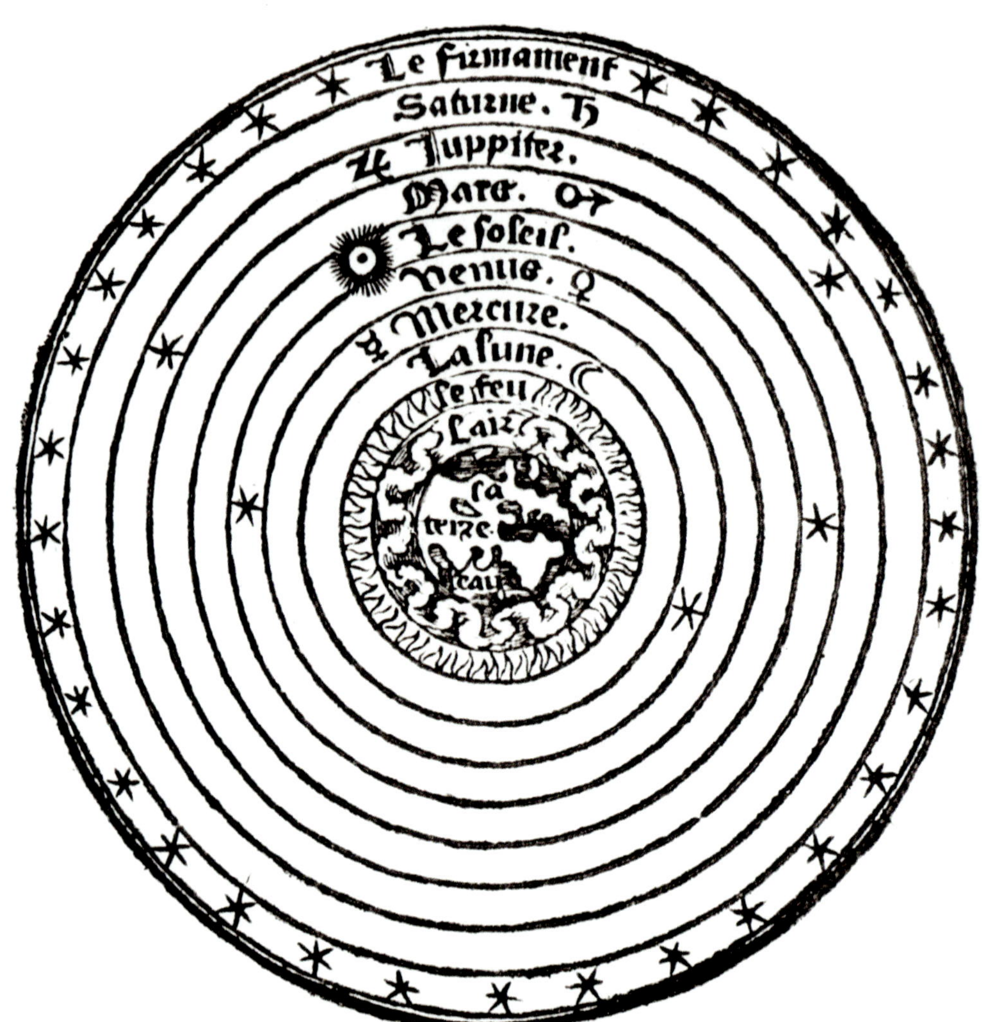

Aristotle's version of the universe, shown here, had some differences from Ptolemy's but was still a relatively simple model.

While these problems might seem like just reason to throw out his model, we have to keep in mind two things. The first is that his model still had a lot of advantages over the competing models. The second is that his whole system of reasoning was held together by the consistency of its logic.

As much as that was a plus, this consistently logical approach was also the foundation of its greatest weakness. Logical arguments like these could not be empirically observed or tested through experiments. To be clear, it would, in theory, be possible to test whether the planets go in circles (though not for Aristotle's day). However, how could you even begin to confirm or reject the claim that a circle is the natural movement of spheres?

Aristotle's system seemed like science but was really based much more on elements of his philosophical theories about the universe. Claims that cannot be tested are detrimental to scientific progress. Such progress requires a mechanism by which we can measure our claims and arguments against the world, form predictions, and then see if they come true. Without that, we have a situation where incorrect theories can persist for centuries because they can't be disproven.

Ptolemy's Theory

This brings us to Ptolemy, whose full Latin name was Claudius Ptolemaeus. Ptolemy was an Egyptian astronomer who lived around 100 CE to 170 CE. Historians are unable to be sure whether his family was ethnically Greek or Egyptian. In either case, his education and his style of thinking was definitely in the tradition of Aristotle and the classical Greek astronomers before him.

We also know very little about his life. Yet the model of the universe that he presented in his great astronomy text—called the *Almagest*—presented an alternate to Aristotle's, and it was to become a pillar of science for the next 1,300 years. His text also included the results of over twenty-five years of careful astronomical observations.

The beginning section of the book contains arguments for the basic elements of his model for the universe. The centerpiece of his model, of course, was Earth, which was a sphere. His argument in support of this took an untestable assumption for its beginning, just as Aristotle's arguments did. All objects fall toward the center of the universe. Since we can test for ourselves that all objects dropped or thrown fall to Earth, we can (incorrectly) conclude that Earth must be the center of the universe.

Even given some of the shortcomings of this approach, Ptolemy did develop some rather ingenious improvements to the Aristotelian model of the universe. Unlike the earlier model, Ptolemy's planets were not fixed within just one sphere. For Ptolemy, only the stars were fixed in place in the largest, outer sphere. Instead, each planet had two spheres. The first was called the **deferent**. The second was called the **epicycle**.

The deferent was a sphere much like those in the model of Aristotle and those who came before him. Then, the second, smaller sphere, the epicycle, had a center located on the edge of the deferent. To help imagine this in your mind, consider any number of amusement park rides where one small group of seats are spinning around on an axis, and there are several such arms connecting to a center, and that center axis is causing all the arms to spin around like spokes on a wheel.

The deferent and the epicycle helped to account for a number of things that did not fit with Aristotle's model. But even then, Ptolemy had to make the model more complicated by moving the deferent slightly so that its center was not shared with the center of Earth. This shifted sphere was called an

This fifteenth-century painting of Ptolemy is in the Louvre's collection.

eccentric. This gets a little harder to picture. For this we have to imagine if the center column of our amusement ride is, itself, traveling in a small circle instead of staying in one place. Our result would be a distinct kind of wobble. While harder to use, the addition of eccentrics helped take care of even more small irregularities. More importantly, the eccentric was an attempt to help explain why the planets might seem to travel at different speeds over time.

In addition to the problem of having some central assumptions that could not be tested, this new model had introduced a new problem. There was no obvious limit for how many wobbles and orbits could be added on. In fact, Ptolemy's successors went on to add eccentrics onto other eccentrics and epicycles onto other epicycles. The result was a more and more complex model in which the various movements of the planets that seem irregular are actually the result of additional, smaller circular movements. We will see the problem with this in more detail in chapter 4.

Because the role of mathematics in astronomy would later represent the heart and soul of the Copernican Revolution, we must credit the work Ptolemy did to add mathematics to the project. In fact, the name *Almagest* was given to it later, basically by taking a Greek word and an Arabic word and smashing them together to mean "the Greatest." Ptolemy's title for the book was *He Mathematike Syntaxis*, which is Greek for "The Mathematical Collection." When taking into account the tables that give **trigonometry** values for different sections of various circles, calculations of the size of the sun and the moon (and their distances from us), and an introduction to the area of spherical trigonometry, it is not hard to see why he named it that.

In the next chapter, we will turn to examine some of the particular problems of the Ptolemaic model and why those

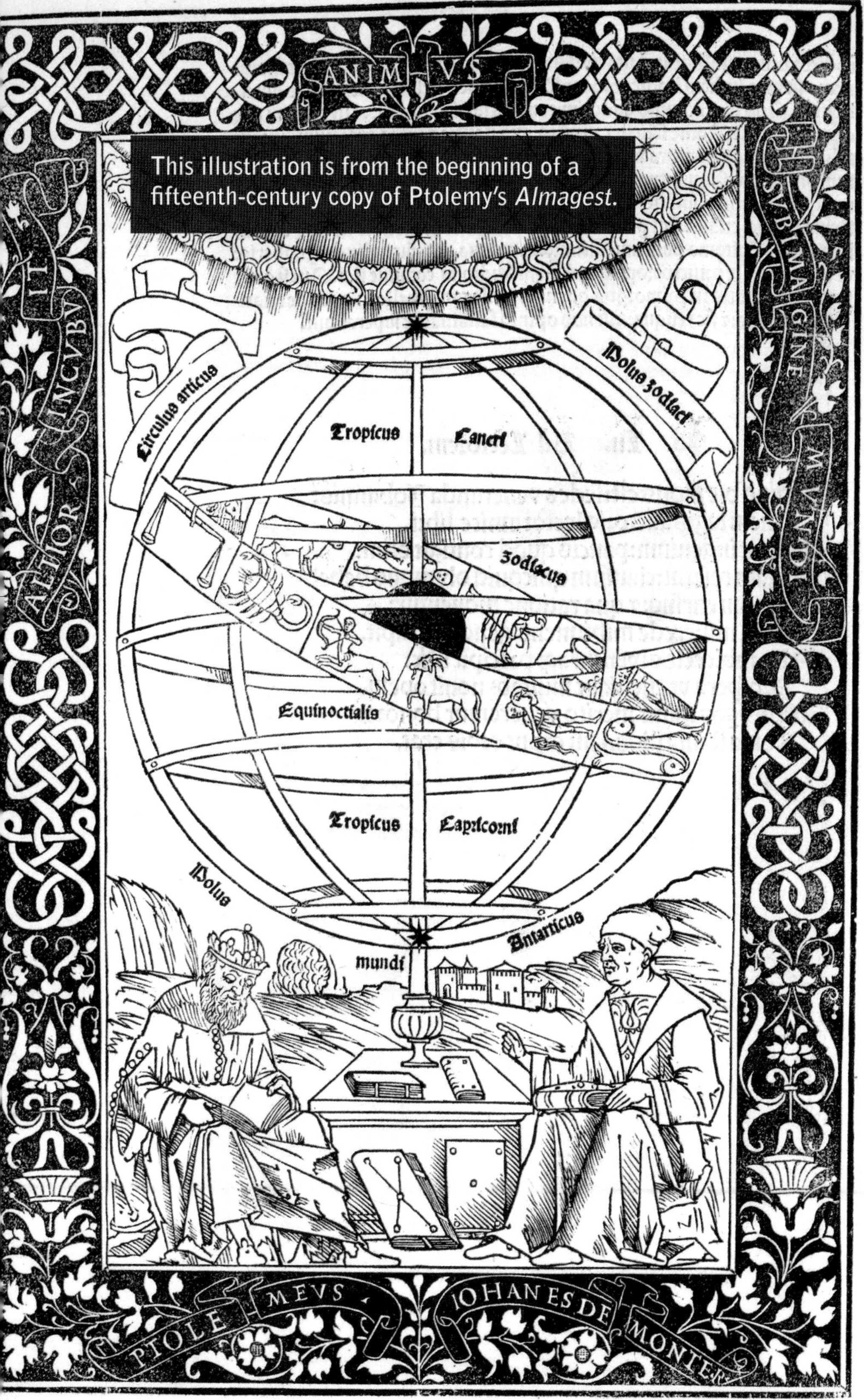

This illustration is from the beginning of a fifteenth-century copy of Ptolemy's *Almagest*.

problems were so important. A few of them we have touched on already, but we need to drill down another layer to see how they are really a closely related group of puzzles to solve. These problems are so hard that it took a handful of the smartest people who ever lived to solve them, and even with that level of brainpower, each of them only solved part of the bigger puzzle. Just as it was suggested at the beginning of the chapter that we should not sell Aristotle and Ptolemy short, this is true of the thinkers to come as well. Without each one taking another step forward, it is not clear any of them could have pulled off the Copernican Revolution.

Schrödinger and an Explanation for Our Scientific Curiosity

In tackling the problem of understanding the universe, it might occur to you, as it has to many thinkers in history, to ask: But why? Why do we work so hard to learn how many stars are in the galaxy next door? Erwin Schrödinger (1887–1961), an Austrian physicist and Nobel Prize winner, was one of the brilliant minds who helped modern science take another leap forward by developing quantum physics, which offered an interesting answer.

Schrödinger thought science was like a game, but this is a game that matters because it is about reality. Therefore, one part of the answer is that we pursue science for many of the same reasons we pursue other games—because the thrill of each small victory is joyful. The excitement of discovering new things combines with the sense of achievement when we have to improve our own abilities in order to unfold those discoveries.

Schrödinger also made the case that God is the other player—he is the one on the other side of the game board, and he is the one who set up the game. Part of the rules we know, but part of them we have to figure out as we go. That is the game, and, for Schrödinger as much as for Kepler or Galileo, part of the reason to play is to glimpse some part of the mind of God. While he would live some two centuries after Newton, you will recognize some of his same sentiments reflected in the discussion of the revolutionaries.

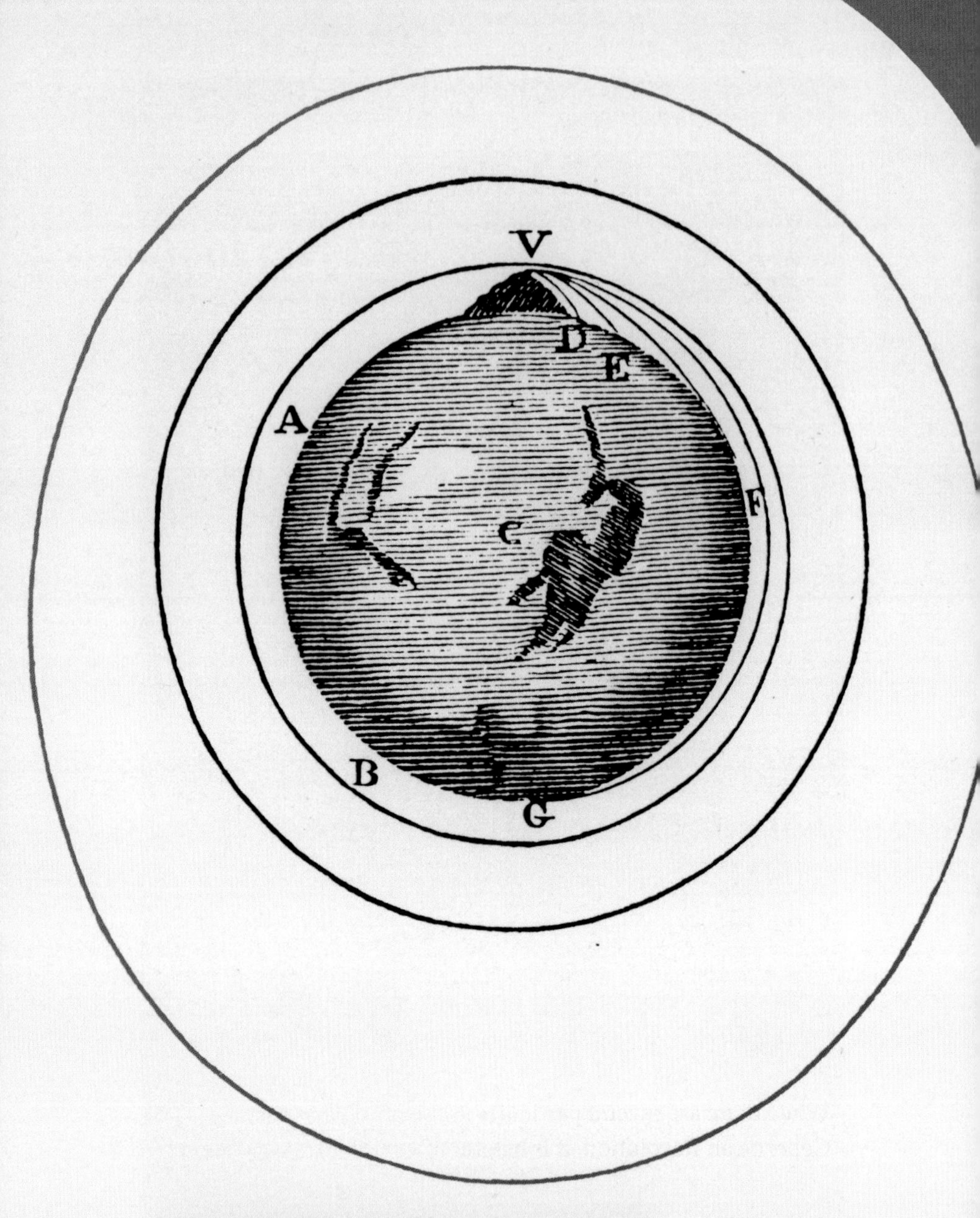

An illustration from Newton's *Principia* shows projectiles moving at different speeds until orbit is achieved.

CHAPTER 2

The Science Behind the Copernican Revolution

In this chapter, we'll look at the specific issues that helped ignite the Copernican Revolution and the events that needed to happen and the tools that needed to be invented in order for it to succeed. As we go, please keep in mind that we must always allow that the story leaves some pieces and some characters out. If not, it would take as long to tell the whole thing as it did to live through it. It will seem, for instance, that no one had any great ideas between the death of Ptolemy and the birth of Copernicus. Yet there were, indeed, several key thinkers and a number of important changes in the way science and math were done during that interval.

The PROBLEM of the PLANETS

While there are several particular problems that drove the Copernican Revolution, it is helpful to focus in on the motion of the planets as the centerpiece. To be more specific, the central problem was coming up with a model that would accurately account for the observed movements of the planets in relation to the sun, Earth, stars, and, of course, each other.

Matejko's 1873 painting portrays Copernicus talking with God.

The term "planets" was taken from the Greek word for "wanderer." From earliest times, observations of the heavens showed that these objects moved around in comparison to the stars, which seemed to stay in fixed locations compared to each other as they rotated around Earth. For the Greeks, the universe consisted of Earth, the stars (counted as one object in most respects), and then seven planets—which meant the sun, the moon, and then Mercury, Venus, Mars, Jupiter, and Saturn.

According to the Greeks, the planets would usually rotate around Earth traveling eastward as they passed in front of the different constellations. (Of course, even Greeks were able to see that they do not hold a constant rate of travel.) This seemed consistent and elegant, and it fit with a very simple model of the solar system for them.

A bigger problem arises, however, because this normal movement is interrupted by brief intervals when each planet seems to eventually slow down and then travel in the wrong direction. Slowing down and speeding up made the problem more complicated, to be sure, but not as much as going backward! When a planet does this, moving westward instead of eastward, it is in **retrograde**. This happens in a regular cycle, though different for each planet. For Mercury, it wanders backward in its path every 116 days. For Mars, it happens every 780 days, and so on.

A second problem comes up, however, in that the planets also seem to behave a little strangely compared to the sun. Two of the planets—Mercury and Venus—travel staying very close to the sun. Mercury, for example, is always within 28° of the sun. In contrast, the other three that we would call planets (Mars, Jupiter, Saturn) vary much more in their closeness to the sun, sometimes even as far as 180°. That would mean each planet was on the opposite side of the circle from the sun at different times.

This, then, sets up a third problem that was hard for the Greek model to make sense of. While the planets all varied in brightness from time to time in their paths, the logic was that this was a result of traveling nearer or farther from Earth. That seems sensible enough. However, when Mars, Jupiter, or Saturn were farthest away from the sun they were at their slowest (the time they went into retrograde) and at their brightest. There was not an easy tool for the Greeks to connect those pieces.

To compensate for these different strange behaviors, then, the ancient astronomical models had to add certain extra features. It made the models more complex, but often it would allow the more complicated model to still be used to account for the motion and location of the heavenly bodies. While a circular model could account for the annual path of the sun and moon in relation to Earth more or less, it could not even begin to explain things like going backward.

It didn't end there, though. Once the model had been twisted up and tangled to fit with the observed backward movements and to account for the irregular travel times around Earth, the astronomers realized there were still a lot of smaller things that could not quite fit into the model—even with the changes they had made. This resulted in a model with extra smaller loops added on to the larger loops. Small circles of movement, for instance, would not be noticeable to our observation as backward travel but would have the effect of slowing progress in different ways.

It is important to note that not all of these adjustments were developed by Ptolemy himself, but because they followed upon his strategy for solution, they are still considered as part of the model developed by him. While subsequent thinkers introduced successive layers of modifications, it was merely a matter of adjusting and fine-tuning the approach rather than a consideration of making some large-scale changes to the model.

Do not misunderstand; while the ability to describe this Ptolemaic model comes across rather simply, the details of it were the result of fairly complex math. Figuring out where the variations occurred, what radius the additional circles would need to be, and so on, helps us appreciate why Ptolemy's greatest work, the *Almagest*, is packed full of diagrams, tables, trigonometry, proofs, and long calculations.

That being said, each increase in the ability of the model to accurately predict the motion and location of the planets was always at the cost of increasing the complexity of the model. Even with many, many tweaks, the model was still only giving an approximate value and could never seem to keep up (or catch up) with improvements in observations. The burbling stream of observations that could not fit in turned to a raging river with the development of the telescope and the Copernican Revolution. But before that could happen, there were a few other key developments that needed to take place.

The RISE of MYSTICAL MATHEMATICS

One of these developments happened as great thinkers attached to the ideas of Plato, rather than Aristotle. The tradition of philosophy, which had influenced a number of important figures and their approach to solving puzzles and questions about the world, was known as Neoplatonism. The Neoplatonists felt that the ultimate form of explanation and research was to study the changing and messy world around us in order to discover the eternal patterns under the surface. The forms of geometry and mathematics were the true knowledge that could only truly exist in the mind. Material things were only ever rough or sloppy attempts to imitate those pure truths in the way triangular things are sometimes close to a perfect triangle, but never quite.

No lines or points that we can ever draw actually match up to the definitions Euclid gives, for example.

The rise in the Neoplatonist philosophy meant that many scientists in the Renaissance and after searched enthusiastically to find patterns of geometry and mathematics in nature. Domenico Maria de Novara, for example, was one of Copernicus's teachers and his friend. Novara was a strong Neoplatonist. He was one of the first to criticize the messy, overly complicated model inherited from Ptolemy. As he described it, nothing so messy and cumbersome could represent the underlying mathematical shape of nature.

Our modern appreciation of astronomy requires very complicated math. Astronomy, at its core, is mathematical. Understanding celestial objects requires an understanding of regular laws and rules of mathematical relationships. This is a far different approach from the endless list of exceptions to the rules that the Ptolemaic model had become.

We can see this line of thinking shine through Copernicus's work in any number of places. His goal of uncovering the mathematical simplicity seems to have motivated many of his ideas and hypotheses. The same sentiment stands out with special clarity in Galileo's famous quote that, "the universe … cannot be understood unless one first learns to comprehend the language in which it is written. It is written in the language of mathematics … without this, one is wandering about in a dark labyrinth." Kepler, too, would tell us, "Before the origin of things, geometry was co-eternal with the Divine mind," and, even better, "The chief aim of all investigations of the external world should be to discover the rational order and harmony which has been imposed on it by God and which He revealed to us in the language of mathematics."

In fact, Kepler is a fascinating case because in several places his enthusiasm for what he thought the mathematical

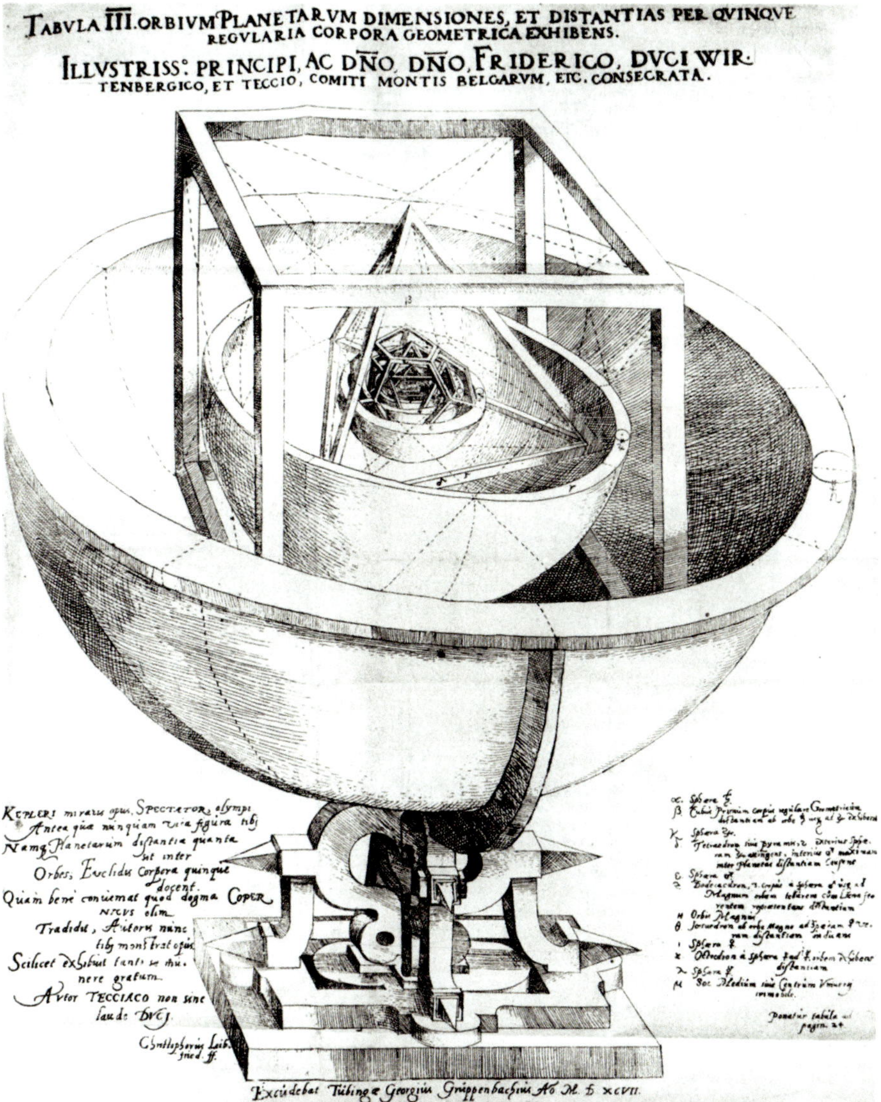

Kepler's model of the universe involved spheres defined by basic geometric solids.

In this image we see the same approach to perspective that revolutionized art in the Renaissance.

pattern *must* be caused him to try to stretch the models and observations a little. Working from the heartfelt faith in the idea that God had written the book of nature in the language of mathematics, Kepler argued in his first major book, *The Cosmographic Mystery* (1596), that the arrangement of the planets fit a particular geometric pattern. Specifically, he thought that if you placed the basic solids in a particular order inside of each other you would get the same spacing. According to Kepler, the order was octahedron, icosahedrons, dodecahedron, tetrahedron, and cube.

In the end the rigor of his contributions was amazing, so we should not let Kepler's overenthusiasm for mathematical ideas cloud our judgment too much.

The RISE of PRACTICAL MATHEMATICS

We can point to a sort of religious vision of math in the work of the Neoplatonists and in their influence on the central figures of the Copernican Revolution. At the same time, there was a second powerful trend in the role of mathematics that had been building from the fourteenth and fifteenth centuries onward. This trend had to do with the ability of applied mathematics to solve any number of real-world problems, from adding realistic perspective to paintings to developing the best designs of military fortresses.

Art

Perspective painting had emerged at the beginning of the 1400s in the work of Filippo Brunelleschi, who was working under the patronage of Lorenzo de Medici, the ruler of Florence, Italy. Ideas about perspective would be developed by Leon Battista Alberti, in 1435, in a text titled *On Painting*. These ideas would find fuller expression in the work of later artists. The heart of this new approach was to train the artist to think in terms of lines and angles from the artist's eye to the objects in the scene. Artists began to start with simple geometric shapes as the skeleton when sketching the underpinnings of the painting.

This, naturally enough, sparked an interest in mathematical angles and shapes and how to understand them in order to use them more fluently in their artwork. As artists' work and study progressed, new ideas arose about empty space—a concept foreign to thinkers like Aristotle. Empty space would take on profound significance in the work of scientists, Newton perhaps most of all. By the mid-1400s, artists' curiosity had also extended to the notion of infinity. This was a key part of thinking about how lines in a painting related to the horizon. Yet the concept was also completely alien to the way people like Aristotle and Ptolemy thought about the universe.

One of the more interesting side effects of this development in art was the ability to represent new inventions. Inventors, tinkerers, and engineers could use pencil and paper to construct pictures of a three-dimensional object. The many inventions that Leonardo da Vinci drew but never built provide a familiar example for us. It also became much easier to share ideas with others by means of such illustrations, rather than having to lug a small machine to them (or wait for them to come to you). A great example of this was the Jesuit astronomer working in China who was able to build astronomical instruments based on drawings—seeing the original objects would have been too great an obstacle.

Cartography and Navigation

The new appreciation for applied math also found a home in the field of cartography (drawing maps). While using geometry was certainly not new to the study and creation of maps, it was

Martin Luther's battle against the Catholic Church would both help and hinder the Copernican Revolution.

taken to a new level in the 1500s. In a book by Peter Apian, for instance, titled *Cosmographic Book* (1524), we see the author transform the sphere of latitude and longitude onto a flat surface. He used geometric perspective, much like that used by the artists. We can imagine this well enough if we call to mind a perspectival drawing of a tiled floor.

Applied math also takes a role in the design of new instruments and methodologies for navigating at sea. Navigation would reach its new apex when Galileo developed a more accurate method of determining a ship's location at sea based on the position of Jupiter's moons.

Ballistics and Fortresses

A surprising example of applied math is found in a work called *New Science* by Italian mathematician Niccolo Fontana Tartaglia. Here "new science" refers to ballistics (the study of projectiles, guns, and so on). The cover page of his book shows Euclid, that brilliant Greek credited as the grandfather of geometry, guarding the gate that leads to knowledge of not just ballistics but knowledge in general. Tartaglia would go on to develop new tools for surveying and new techniques for calculating how far away a particular target was. (More than one important battle in those days was decided by a single successful hit on a key target.)

We might not think of fields like painting or ballistics as part of the Scientific Revolution, or even, often, as sciences at all. However, we can see that they played an important role in helping to shape the way people tackled new problems. After all, these problems were not just in the heavens above, but in the world around them as well.

The impulse to tackle real-world problems characterizes the key figures in the Copernican Revolution, too. Copernicus did not just work on astronomy, but on monetary reform and small inventions. Similarly, Kepler not only solved the puzzle of changing speeds of planets (which we'll discuss later) but also did some interesting work on the volume of wine barrels and on the most efficient way to stack cannonballs. Galileo's funding, which allowed him to improve upon the design of the telescope, came from the Venetian government because of its potential military applications. (On the open sea, whoever can spot the opponent's navy ship first has a big advantage.)

Naturally enough, this same application of mathematics also took on new significance in the design and placement of military fortifications. Commanders wanted to know with improved certainty where to place cannons for optimum range, but also where to place them to prevent being taken out.

TEARS and TRANSFORMATIONS in the SOCIAL FABRIC

Around the time of the Copernican Revolution the church still held sway over most of the universities. However, the new emergence of independent nation-states provided new placements and patronage arrangements for mathematically oriented scientists. Of course, patrons were eager in no small part due to the various applied uses of the new mathematical approaches, as discussed above. Nobles needed strong ballistics and navigational experts on hand as they fought to protect their land and expanded their holdings.

At the same time, the other major upheaval underway was the **Protestant Reformation**. Of course, this movement resulted in a crackdown on independent thinkers in a way (as evidenced in what seems to be an overreaction to cases from Bruno to Galileo by the Catholic Church). Yet it also created some breathing room for scholars and scientists who were not willing or able to carry on their research under the rigid priorities of the Catholic hierarchy. We will see this touched upon briefly in looking at some of the biographical details of major characters in chapter 3, and perhaps most importantly, we see those forces at work in the person of Georg Joachim Rheticus, who was an indispensible figure in helping to launch the Copernican Revolution in the first place!

The RISE of INFINITE MATHEMATICS

While we have largely been discussing the forces and changes that took place leading into the Copernican Revolution, we must take a moment to reflect on what ended up being one of the most significant inventions of all: **calculus**!

Calculus did not come along until the Copernican Revolution was well underway, but without it, the whole enterprise would have been only slightly more developed than the Ptolemaic model it had replaced.

As part of the mathematical need for understanding ballistics and laws of motion in general, Galileo ran into some problems. Without being able to figure out how fast an object was traveling at different times, or how its direction was changing at each point along the way, it was impossible to form predictions that were better than rough guesswork through trial and error. Guesswork and trial and error could never explain things like the motion of planets or, more immediately, the behavior of cannonballs fired at a distant fortress.

Being a good scientist and mathematician, Galileo wanted to formulate a law for how bodies fall. He began his work with the simplest possible path: falling straight down. Based on his experiments conducted at the Leaning Tower of Pisa, he was able to show that if an object falls for t seconds (a function of time), and the **gravity** acting on it stays the same, then we could try to shrink the distance being measured until it was almost zero. That would allow us to figure out how fast it was traveling at a given instant (traveling almost zero distance takes almost zero time).

These theories became more impressive once Isaac Newton (from England) and Gottfried Leibniz (of Germany) began making contributions. Even at this point, though, we can appreciate how many ideas have come into play that would have been completely bewildering to Aristotle and Ptolemy: zero, infinitely small distances (or times), and the idea that planets follow the same rules as things dropped from the Leaning Tower of Pisa.

This kind of mathematical thinking started the process in motion of connecting geometry and physics. Newton and Leibniz were working at the same time as one another to push this connection to the next level.

The two men wanted to establish some simple rules that could be used for finding the formula for the **slope** of the line that was **tangent** to a curve at any point on it, given only a formula for the curve. That means, basically, they were looking for the specific direction the object was traveling at any particular instant. This is not so puzzling for the case of dropped things, of course, but very important for planets or cannonballs.

Each of them was working on this problem without being aware the other one was doing the same kind of math. They also had no idea that they would actually change the world with their brilliance. The process they came up with for figuring out the slope of tangent lines for any given point on a curve was called differentiation, which is where the name "differential calculus" came from.

One of the important applications of differential calculus came when they worked further and figured out how to go in reverse. Based on the idea of summing up the areas of an infinite number of rectangles drawn under the curve, with each of them being infinitely thin, and then undifferentiating the result (called integrating), they could find the distance traveled by a given thing if they knew its **velocity**. Again, notice that we have had to introduce infinitely small rectangles, and infinitely many of them, into the thinking process in order for this new mathematical invention to work.

The Power of the New Calculus

Calculus's role in making the Copernican Revolution a development of lasting value cannot be overstated. By means of this new calculus, Newton was able to figure out several unsolved puzzles regarding gravity, as will be touched on in chapter 4. But it also allowed him to translate Kepler's brilliant work on how the planets moved into more consistent mathematical terms and fit them together in one larger model with other developments.

Two incidents in particular, though, help to highlight how powerful this new invention was. The first of these was the development of the book *Celestial Mechanics*. This massive project was twenty-five volumes long and published over the years 1799–1825. Pierre-Simon Laplace, the author, was a brilliant mathematician in his own right. This book translated all of the work Newton achieved in fields of motion and physics into the framework of calculus. In this book, Laplace was also able to use calculus to help correct some unsolved problems concerning what seemed like irregular planetary orbits in Newton's model. An example of this is how the conflicting gravitational pull of the sun and Earth cause the moon not to follow a clean elliptical path, as would otherwise be expected.

The second incident to point to is the discovery of Neptune. Urbain Jean Joseph Le Verrier, a French mathematician and astronomer, was working to solve a discrepancy in the orbit of Uranus. Its observed positions were slightly off compared to the model Newton had built. Once Le Verrier worked out the calculations, which indicated the existence of an additional planet, he was able to tell astronomers where to look for it. German astronomer Johann Gottfried Galle found it within an hour (and it was only 1° away from where Le Verrier predicted, which is remarkably accurate). This was September 23, 1846, and the incident was considered one of the greatest moments of nineteenth-century science. It's not an exaggeration to say it ranks right up there as one of the single most amazing moments in science *ever*. As someone quipped, Le Verrier had discovered a new planet using just a pencil and paper!

Tyco Brahe's Notebooks

Tycho Brahe was a Danish astronomer who lived from 1546 to 1601. While he did not make any of the major breakthroughs central to the Copernican Revolution, he is a great illustration of the other side of the science coin. His patient observations and careful collection of data were indispensable resources that Kepler and others used to accomplish so much.

When he was studying law at the University of Copenhagen, Brahe witnessed a total solar eclipse, and it made a powerful impact on him. From that point on, Brahe continued to study law during the day, but each night he spent learning about the stars. In 1563, he observed an overlap between Jupiter and Saturn. When he found that the existing books that listed the positions of stars and planets were several days off, he set out to make his own measurements in order to update the reference materials.

Brahe spent the next five to seven years traveling around Europe to learn more about astronomy and gather the instruments he would need. He then settled down and set up a small observatory. That was soon replaced by the establishment of a larger observatory, which was made possible when the king of Denmark gave him a small island and financial support for the project.

He was the last major astronomer to work without a telescope, and this fact makes it all the more amazing that his observations were almost five times more accurate than the ones his contemporaries had been using.

A Turkish manuscript depicts the comet of 1577 that would play a part in the work of Tycho Brahe.

Copernicus is honored by a statue in his homeland of Poland.

CHAPTER 3

The Major Players in the Discovery

To understand a major historical event or period, it helps to think of it sometimes as a play being performed in the largest theater possible. When you watch a play with a complicated plot, it also helps to have a list of characters with a brief overview of each one. That is key to keeping track of the different threads of the story, and it is the only way you can really appreciate the different twists and turns that take place. Let us turn, then, to the introduction of the main characters in the story at hand.

ARISTARCHUS

The first person responsible for what would come to be known as the Copernican Revolution actually lived nineteen centuries before Copernicus was even born. Aristarchus of Samos is usually just referred to as Aristarchus, but since there are one or two other people with famous achievements and the same name, including where he was from helps to clarify.

Aristarchus was born in Greece—Samos is a small Greek island just off the west coast of the country—around 310 BCE and died around 230 BCE. During his lifetime, Aristarchus was

Aristarchus lived from 310 BCE to 230 BCE.

most famous for developing ways to use geometry to figure out the size of the sun and the moon and their distance from Earth. His line of reasoning was amazing, and a couple of astronomers who came after him tweaked his calculations to get even better measurements, and these measurements were incredibly accurate for their time.

Aristarchus's other great scientific project was writing a book that appears to be the first to build a case for heliocentrism. Although the book didn't survive, his work was mentioned by some other famous thinkers, like Archimedes. Aristarchus's idea that the sun is at the center of the universe was generally disregarded in favor of the geocentric models of Aristotle and Ptolemy; however, Copernicus gave Aristarchus credit for having sparked his own thinking and research on the topic. An early draft of Copernicus's famous book *Six Books Concerning the Revolutions of the Heavenly Orbs* is even dedicated to Aristarchus, though ultimately the tribute was not included in the final published version of the book.

COPERNICUS

Although Copernicus was not the first to imagine the sun as the center of the universe, he put the idea on solid scientific footing and thus launched not only new models of astronomy, but new ways of approaching and advancing all of science.

Nicolaus Copernicus was born Mikolaj Kopernik in Prussia (now part of Poland) in 1473; he is best known today by the Latin version of his name. Copernicus was raised in a thriving seaport town where his father was a successful merchant. Copernicus's mother was also from a prominent merchant family, and he was the youngest of her four children. His older brother became a priest, one older sister became a nun, and his other sister married and had a big family. While

Copernicus never had children of his own, he became the guardian of his sister's five children in his later years.

Through his mother's family, Copernicus was connected to many other wealthy families and some Polish nobles. His father died when Copernicus was about ten, and one of these prominent relatives became a patron, helping him to get a good education and to launch his career. Although he would later go on to study in other fields, including medicine and church law, it was during this education at the University of Krakow that he gained such a thorough understanding of mathematics. It was also where he learned about the astronomy of Aristotle and of Ptolemy—and the flaws in and differences between their cosmological models.

Even with Copernicus's understanding that heliocentrism solved the problems seen in the work of other astronomers, there were some gaps in his understanding, so he kept pushing back the publication of his greatest work until he was dying, in 1543. One such gap is that he was still trying to use the geometry of circles to calculate the different positions of the planets in relation to each other. Adding in some smaller circles could help make the observations match up, but this ultimately resulted in more and more exceptions to the model. Other genius astronomers and scientists would complete the revolution he started.

GEORG RHETICUS

In 1539, Copernicus was joined in his work by Georg Joachim Rheticus, a twenty-five-year-old professor of mathematics and astronomy from the University of Wittenberg. He had been sent by a prominent theologian (an ally of Martin Luther, the religious reformer) to study with leading astronomers. Rheticus ended up staying with Copernicus for two years, studying with him and helping with his work.

Rheticus was born as Georg Iserin in Feldkirch, now part of Austria, in 1514, and he died in 1574. Rheticus's parents were wealthy, and his father had clout as the town doctor. Unfortunately, Rheticus's father abused his patients and stole from them. As a result, he was tried and executed in 1528 (when Georg was about fourteen). Because his family was stripped of his father's name, Georg and his mother went by her maiden name of de Porris. Georg later gave himself the last name of Rheticus when he was at college to reflect the region of his hometown (similar to the naming convention of Aristarchus of Samos).

Rheticus is sometimes referred to as the first Copernican because working closely with Copernicus on the refinement of his ideas made him the first to really see the scope and value of heliocentrism. Rheticus collaborated with Copernicus to publish a shorter version of the new ideas in a book called *First Narration* (in 1541), and Rheticus was listed as the sole author in order to test out reception to Copernican ideas. To help the book seem more appealing and more acceptable, it was also put into the context of helping researchers and scientists derive better, more accurate calculations for planetary orbits and locations.

Rheticus later went on to publish his own large work in the area of trigonometry, though he is mostly remembered for his work with Copernicus. If not for the tireless efforts of Rheticus, Copernicus's great work might never have made it into publication.

GALILEO

Galileo Galilei was born in Pisa, Italy, in 1564, and before he passed away in 1642, he would manage to make huge strides forward in the study of motion, gravity, astronomy, mathematics, philosophy, and even the scientific method.

Galileo was the oldest of six, though two of his siblings died as infants. His father, Vincenzo, was a musician who liked to experiment and made some contributions to the study of music theory. After attending a monastery school, Galileo attended the University of Pisa to study medicine in 1581. While there, he ended up in a seminar on geometry by accident, and he fell in love with mathematics. Galileo convinced his father to let him change his major from medicine to math, and the whole course of history changed when his father agreed.

Galileo left the university in 1585 without finishing his degree and spent the next few years giving private math lessons. During this period, he first came to the attention of the scientific community when he invented a special kind of balance used for measuring tiny stuff (like precious metals), which he wrote a small book about.

From such humble beginnings, Galileo rose to become one of the greatest minds the world has known. At the heart of his many pursuits was the struggle to understand the universe.

Galileo encountered resistance to his findings, and his commitment to the scientific method would have profound consequences, including a contentious relationship with the Catholic Church that led to his imprisonment. The controversy started because certain passages in the Bible say, for example, that God set the Earth on its foundations and it cannot move. In defending his theory, Galileo wrote a letter arguing that the Bible is a great authority on things like faith and morality, but is not intended to be a textbook of science. This sounded to the members of the Inquisition like the reinterpretation of the Bible during the Protestant Reformation. It should be mentioned in fairness that the church did rely on arguments from other prominent scientists, like Tycho Brahe, in reaching its judgment, but in

Galileo is shown here demonstrating one of his designs for the telescope.

the end, the church ordered Galileo to deny the truth of his research regarding heliocentrism.

The whole escalation of the issue was perhaps brought to an extreme because Galileo had insulted a number of Jesuit scientists a few years earlier. As part of the controversy with a Jesuit astronomer over the nature of comets, Galileo ended up publishing a small book, *The Assayer*, that dismantled the astronomer's arguments. Many historians believe that this text and the arguments set out helped influence the Inquisition to pass a harsher sentence on Galileo than they otherwise might have.

Even after the church forced him to recant many of his writings, demanded he stop making arguments for the heliocentric model, and put him under house arrest, Galileo continued to push science forward.

In the face of these personal obstacles, Galileo's work had a major impact. Some of his experiments revealed flaws in other Aristotelian ideas. Galileo also laid some of the groundwork for work by René Descartes and Newton concerning uniform motion and gravity. In fact, part of Galileo's work would help Albert Einstein figure out elements of **relativity**.

JOHANNES KEPLER

Johannes Kepler lived and worked at the same time as Galileo. Kepler's brilliant work would tilt the scales and guarantee the success of the revolution that Mikolaj Kopernik had started.

Kepler was born in 1571 in Württemburg, Germany. While Kepler's grandfather had been the mayor of the city, their family fortunes were shrinking, and Kepler's father made a living as a mercenary. His mother was the daughter of an innkeeper. Kepler's father abandoned the family when the boy was about five. He suffered from smallpox as a child, which left him with poor vision and limited use of his hands, making the observational research of astronomy he would undertake even more challenging. At the same time his body struggled, he had several moments of inspiration that would help set him on his eventual path. One of the most striking of these was witnessing a comet when he was six and a lunar eclipse when he was about nine.

Kepler studied philosophy and theology at the University of Tübingen, but mathematics was his true gift. Even as a child, his mathematical abilities had amazed travelers who stayed at his grandfather's inn. It should come as no surprise, then, to find that he mastered both the Aristotelian and the Copernican theories of astronomy (which were still part of the math departments, rather than physics departments, at the time). Among his teachers was the most highly respected astronomer in Germany, and this teacher was quiet about being one of the rare Copernicans. His influence explains how Kepler became so well-versed in both approaches. Kepler did so well at his university that his professors recommended him for a position as a math and astronomy teacher when he graduated.

Kepler's first major project was publishing a text defending the Copernican theory called *The Cosmographic Mystery* (1596). Not only was it his first major published work, it was also the first defense of Copernicus published by anyone. For Kepler, astronomy brought together his love of math and theology. Part of the wonder and fascination was that he saw mathematics as unfolding the secret of God's design of the universe. While that got in the way of his scientific research sometimes, it was balanced out by the passion and motivation that were stoked by this connection between science and faith.

Kepler became a mathematics teacher at a seminary school in Austria, and he would later go on to become the assistant to the noted astronomer Tycho Brahe. Eventually, he became the imperial mathematician to three emperors. In addition to that, Kepler's work on **optics**, along with an improved design for a **refracting** telescope, led to mentions in some of Galileo's work.

Newton was born in a quaint country setting at Woolsthorpe Manor.

When Kepler solved the puzzle of planetary orbits by switching from the use of circles to ellipses, the balance of opinion among scientists shifted further in favor of the Copernican model compared to the older ones. Then when he added on discoveries for understanding the laws of planetary motion, Kepler's genius had done the empirical work that would set the stage for the English giant of modern science, Isaac Newton.

In his last years, Kepler suffered much misfortune, including his mother being accused of witchcraft, which led to a long and costly trial. She was finally declared innocent, but the many months of imprisonment had damaged her health and she died shortly afterward. Kepler struggled to further his career amid continuing hostilities between religious factions, in part because he had worked to be friends to people on both sides. Then, in the course of trying to recover some investments, he ended up catching a fever. He died in 1630, and although the location of his grave has been lost, we still know the epitaph he wrote for himself:

> I used to measure the heavens.
> Now I shall measure the shadows of the earth.
> Though my soul was from Heaven,
> The shadow of my body lies here.

ISAAC NEWTON

Isaac Newton was born on Christmas Day, 1642, in Woolsthorpe, England, and died in 1727, in London. His father had died a few months before he was born, and by the time he was two, his mother had remarried and moved to a different town, leaving him in the care of his grandmother. His mother soon became

This statue of Newton is at Trinity College, Cambridge, where he studied.

a widow again, so she pulled him out of school and tasked him with farming and managing the land she owned. It was a disaster; young Newton hated farming.

A local teacher convinced her to let him return to school, and by 1661, Newton was ready to enter Cambridge University. In spite of the great work by thinkers like Galileo and Kepler in the wake of Copernicus, Cambridge still focused on Aristotle as if the Scientific Revolution had never started. While studying Aristotle for his courses, Newton began reading the more modern thinkers like Kepler and Descartes.

Newton had also begun to develop one of the most brilliant mathematical minds in history, including tackling the slow process of inventing calculus. (Gottfried Wilhelm Leibniz would also receive credit for the invention; both Newton and Leibniz had been working on it, unaware of each other's ideas.)

Newton graduated with a bachelor's degree in 1665, but he had already done more work on his own than he did for the degree. Then, before Newton could pursue further studies, Cambridge University closed due to the plague. Back home in Woolsthorpe, Newton spent the next two years developing theories about gravity and optics, and he worked to refine his calculus. In 1667, he returned to Cambridge and so impressed the faculty with his work that one year after receiving his master's degree, he was made a professor.

By that year, 1669, even though only a small circle of people had read his short book on mathematical ideas (called *On Analysis by Infinite Series*), he had already become one of the greatest mathematicians in the world. This, really, then became the foundation of his contribution to the Copernican Revolution. Newton would go on to build science upon a rigorous mathematical framework, improving science's capacity to explain natural phenomena. Kepler's laws of planetary motion are just one example: Newton was able to show that they could be derived mathematically.

Newton's masterpiece, *The Mathematical Principles of Natural Philosophy* ("natural philosophy" was what we now call "science"), was published in 1687. Among Newton's many other accomplishments, he would go on to pen works on religious topics and make further developments in the field of mathematics. One of the more colorful roles over the course of his life was to serve as the master of the mint. During his work there, he personally pursued and prosecuted almost thirty counterfeiters. Some of his undercover detective work for this would be deserving of a movie. He received what would amount to $200,000 to $300,000 a year for this job, and when added to his other income sources, it resulted in a wealthy estate.

Though he had been excused from the requirement that Cambridge professors be priests, Newton nevertheless passed away having never married or started a family. He was only the second scientist ever to be knighted (Francis Bacon was the first), and he presided over science in England for twenty-four years as the head of the Royal Society. To appreciate his place in English hearts and world history, we must reflect on the fact that he is buried in Westminster Cathedral, which also happens to be the traditional resting place of the kings and queens of England up until the nineteenth century.

EDMOND HALLEY

Edmond Halley, the English astronomer, mathematician, and inventor, was born as the son of a wealthy soap maker in a small town near London in 1656 and died in 1742. While Halley did not have quite as large a role in pushing the Copernican Revolution forward as some of the others, he stands with Rheticus as vital all the same. Halley's work provides a fitting kind of capstone to the lines of thinking at the heart of that revolution.

Halley worked with Newton in developing theories of gravity and on understanding the forces that would keep planets

Halley's comet played an interesting role in confirming the merits of the Copernican Revolution.

in motion. He also played an important role in editing Newton's great masterpiece, the *Principia*, and getting it published.

Using the laws Newton had developed, Halley—a brilliant astronomer in his own right—predicted in 1705 that a comet that had last appeared in 1682 would make its next appearance in 1758. Halley's work demonstrates the relationship between mathematical formulae and empirical testing of scientific theories. When what has now come to be known as Halley's comet made its sparkling reappearance on Christmas in 1758, it was a night to celebrate, though Halley himself had been dead for sixteen years and would not see the confirmation of his theories. Gone forever were ideas like Aristotle's that the planets were held in concentric crystal spheres, as beautiful an image as that might have been to imagine. Never again would arguments in favor of circular orbits be made simply because circles are "the most perfect shape." Mechanistic, scientific explanations were here to stay.

Galileo Is Condemned

The following excerpts are from the pope's official condemnation of Galileo in 1633:

> We say, pronounce, sentence, and declare that you, the said Galileo, by reason of the matters adduced in trial, and by you confessed as above, have rendered yourself ... suspected of heresy, namely, of having believed and held the doctrine ... that the Sun is the center of the world and does not move from east to west and that the Earth moves and is not the center of the world ... we are content that you be absolved, provided that, first, with a sincere heart and unfeigned faith, you abjure, curse, and detest before use the aforesaid errors and heresies ...
>
> And in order that this your grave and pernicious error and transgression may not remain altogether unpunished and that you may be more cautious in the future and an example to others ... we ordain that the book of the "Dialogues of Galileo Galilei" be prohibited by public edict.
>
> We condemn you to the formal prison of this Holy office during our pleasure, and by way of salutary penance we enjoin that for three years to come you repeat once a week at the seven penitential Psalms. Reserving to ourselves liberty to moderate, commute or take off, in whole or in part, the aforesaid penalties and penance.
>
> And so we say, pronounce, sentence, declare, ordain, and reserve in this and in any other better way and form which we can and may rightfully employ.

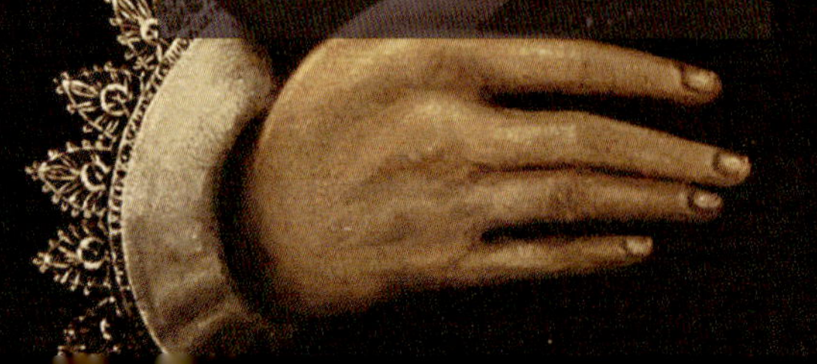

A rather fashionable portrayal of Johannes Kepler

CHAPTER 4

The Discovery Itself

In order to fully appreciate the impact of the Copernican Revolution, we have to look past the surface. Of course, we will talk about the amazing discoveries that came out of the work of the brilliant scientists involved. Yet part of what is important is understanding how their discoveries changed the way people see the universe. The revolution had such wide-ranging implications that we must see this set of discoveries for how monumental they were.

In fact, when the Polish astronomer Copernicus presented his case for the idea that Earth actually travels around the sun instead of the sun traveling around Earth, it made people question their perspective. It seemed to them that the way they'd thought about the world was an optical illusion.

This feeling amplified, then, when the Italian astronomer Galileo used his telescope to show people that their normal way of seeing things like the moon and stars was not accurate either. When he combined that with careful mathematical reasoning, people were able to see that rudimentary observation actually got in the way of understanding them.

There's more, though. Not only did the work of Copernicus and Galileo cause people to suddenly doubt their senses and

Galileo was forced several times to defend his ideas against the church.

what they thought they knew, it also contradicted the Bible. It is hard for many of us to grasp just how the teachings of the Bible were applied to many facets of life and how contradicting its teachings could give way to a kind of mental and emotional dizziness. According to the Bible, the most trusted source anyone knew, Earth was supposed to be at the center of everything in the universe, but there was more and more evidence that the geocentric model just wasn't true.

Just by Copernicus having the courage to move the sun into the center of the solar system, and thus the center of the universe (at least for the people of his day), everyone had their world sort of turned upside down—almost literally. People's sensory observations about things? Undependable. Their trust in the Bible to understand science as well as religion? Undependable. In fact, even the most trusted scientist in all of history up to that point, Aristotle, was also now undependable.

One after another, Galileo uncovered mountains on the moon, other moons circling Jupiter (everyone thought ours was the only one), and **sunspots** on the surface of the sun. In fact, the Milky Way itself was made up of countless stars that no one had ever even suspected were out there beyond the ones we could see with our own eyes in the night sky. All of these discoveries shook people's core understanding of Earth. And the more astronomers learned, the more questions arose.

COPERNICUS'S GROUNDBREAKING IDEAS

The first of these big breakthroughs came from Copernicus. We can almost imagine him there, in the year 1543, lying in bed where he was just about to die, holding his awesome book with a big title. It was called

De revolutionibus orbium coelestium libri VI, which was Latin for "Six Books Concerning the Revolutions of the Heavenly Orbs." Though he would not see it, this book was the first volley in a revolution that would change the world. In fact, in important ways, the revolution is still going almost five hundred years later. The next moment in science that would be this mind blowing would not come until the 1920s when the great astronomer Edwin Hubble proved that there are other galaxies besides the Milky Way.

Even though Copernicus gave credit for his work to a fictional character named Hermes Trismegistus, who was thought of as a kind of mystical idol among the Neoplatonists, the book was a serious piece of science based on observations and mathematics. With this one grand gesture, Copernicus had swept a thousand years of tiny adjustments to Ptolemy's model off the table like pieces on a game board. In its place he put something elegant. The way planets seemed to travel backward sometimes could be explained by the difference between Earth's movement and theirs. Changes in the brightness of planets could also be explained by their movement compared to the sun. Even more, the puzzle about why Mercury and Venus always seemed to stay close to the sun while traveling was solved when Copernicus showed that their orbits were closest to the sun. Of course, the model also had a great appeal to the Neoplatonists because of its mathematical simplicity compared to the messy complex model the scientists were trying to use before.

Now we should not expect Copernicus to solve all the puzzles in one fell swoop. Even as brilliant as he was, his model still used the idea of regular circles for the orbits of the planets. This meant that in some ways his model would have worse predictions than the Ptolemaic one. Thus, he still had to use some of the kinds of tweaks and adjustments as before, but this version of cosmology had made huge strides forward. Copernicus's model was *not* so amazing that everyone who saw it had to buy in completely; there was work left for the scientists who would come after. One of the most puzzling problems with Copernicus's explanations, for instance, was that if Earth was traveling through space, and if the cycle of night and day was due to the spinning of our planet instead of the movement of the sun, what kept everyone from being flung off into space like a playground merry-go-round?

In addition, there was a problem with wrapping our minds around the size of the universe. One of the things that would follow, given this new model, is that as Earth revolves around the sun, the stars should look like they are moving as we go around (like how the scenery changes when you're going around on an amusement park ride). Since Copernicus and others could not detect any shift in the star scenery, then there seemed only two possible explanations. One was that Earth was at the center after all. The other was that the stars were so incredibly far away that they would not seem to shift.

As you might expect, Copernicus went with the second option. However, that meant everyone had to stretch their ideas about the universe to allow for how big it would have to be. They also had to realize that a lot of the universe seemed to be empty space. Not only was that strange for them, but it also created a mystery as to why God would create a universe with so much empty room in it. Why not just make it the size we needed? This was about as confusing as if someone bought you shoes that were ten sizes too big.

It was because of some of the implications like this that many scientists did not embrace Copernicus's new model. Many astronomers knew about Copernicus's work, but they kept teaching about Aristotle and Ptolemy at first.

GALILEO'S CONTRIBUTIONS

After Copernicus came Galileo. He was not one of the astronomers willing to let Copernicus's model go unnoticed. As for the problem of why everything doesn't just fly off Earth, Galileo solved that using logic. Things do not get flung off Earth because in terms of rotations per minute, nothing is really being spun that fast. It also seemed that if Earth were spinning, then when you drop something from the top of a tall tower, the tower will keep moving east, as Earth rotates, but the object is no longer being carried along, since you have let go of it, so it should fall somewhere behind you. Consider that if you are spinning and you drop a ball, it doesn't spin with you. Though we see that dropped things from a height do not fall behind, that is not the same as being able to *explain* it.

Galileo argued that the fact that objects are actually spinning rather slowly explains this puzzle as well. Because there is no friction (well, very little) both the object dropped and the tower are still moving around Earth at the same speed. Galileo argued that moving bodies will keep going the same direction forever as long as there is no friction. His reasoning for this was based on his experiments of rolling balls down **inclined planes**. Observations showed that if one plane was tilted down and one plane was tilted up, both at the same angle of elevation, then rolling a ball down the one would result in it trying to regain the same height by rolling up the other. From this Galileo reasoned that if the second inclined plane were not there (and no surface friction), the ball would, instead, just keep rolling forever.

This is not a bad starting point if you were trying to derive some of Newton's later work. According to Newton, that is the same reason that the planets, once they started going in circles at creation, would continue going in circles forever around the sun. Interestingly, because of this idea, Galileo would never buy into Kepler's model of elliptical orbits for the planets because it couldn't explain why they would keep going indefinitely.

That was nothing compared to the other stuff Galileo discovered. Although he did not invent the telescope, working from a scientific hunch, he improved on the design of existing telescopes with the expectation that there would be some amazing things discovered once he pointed one into the night sky. In fact, he refined telescopes to achieve amplification of 20x (twenty times normal eyesight), which would only be useful for astronomy. The other primary use for telescopes at the time was for ships at sea, but because of the curvature of the horizon, any ship far away enough to need that much magnification to see would be "around the corner," so to speak.

Stars, Stars Everywhere

The first thing that happened when Galileo began pointing his telescope up at the night sky was that he found more and more stars everywhere he looked. The number of stars within the area of different constellations was much larger than anyone thought before. What is more, the Milky Way, which was a kind of blur to the naked eye, was believed to be some different type of phenomenon. Galileo's telescope, however, showed that it was a giant collection of stars.

Mountains on the Moon

In 1609, then, Galileo pointed his telescope at the moon. When he did so, he was able to see so much more than just patches of darkness on the surface, as people had with their own eyes for centuries. Instead, he was able to see that along the edge between the light and dark part of the moon, the

The four largest moons of Jupiter are still referred to as the Galilean moons.

surface was jagged. If it were smooth, as had been thought and taught, this line should be straight and smooth. Instead, there were dark areas crossing over into the light section and light areas where he should have only seen dark. He was able to build the argument that this was the same effect as the light and shadows of a mountain range on Earth when the sun was coming up. The tall parts catch light and cast shadows that stretch out away from the mountains (making darkness extend into the light area).

His careful examinations also allowed him to discover the existence of craters on the moon. He was even able to calculate the depths of some of them (as well as the heights of the mountains).

Moons Around Jupiter

Then, in January of 1610, Galileo pointed his telescope at Jupiter and found that there were moons orbiting it. But he did not gain this insight right away. At first what he discovered were four small lights near Jupiter. As he observed them from night to night, they kept rearranging their positions. Of course this wouldn't make sense if they were stars because no observations had yet suggested that stars move around at all. It also wouldn't make sense if they were planets, for they had such incredibly small cycles of orbit and seemed to follow neither the old pattern of orbiting Earth nor the new pattern of orbiting the sun. His best explanation for this behavior, then, was that they must be orbiting Jupiter.

We now know Jupiter has many moons, but the four largest ones, the ones he could detect, are still referred to as the Galilean moons. They are Io, Europa, Ganymede, and Callisto. In the Ptolemaic model of the universe, all the other heavenly bodies were supposed to orbit around Earth. This also helped show that Earth's moon revolving around us was not a strange exception but seemed to be part of a larger pattern.

Part of what is amazing about this particular discovery is that it did more than just enhance our model of planets, moons, and other celestial objects. This discovery of Jupiter's moons had also, for the first time, identified something in space where it was thought there was nothing.

Phases of Venus

Galileo's next challenge was to untangle the brightness of Venus. Some of the differences in brightness (it was sometimes more bright and sometimes less bright) could be explained by its nearness to Earth—sometimes it is closer and sometimes farther from Earth. However, this didn't quite explain all the variation. For instance, on occasions when the model would predict that Venus should be closer to us, it was dimmer. Similarly, sometimes when the model said it should be far away from Earth, it would seem brighter. Galileo's hypothesis was that Venus had phases just like we observe about the moon. His observational evidence in support helped to prove that Venus did, indeed, orbit the sun as Copernicus's model claimed. This was an important step for the heliocentric model. While we are used to these ideas, we should try to imagine how amazing it was to test it out and collect evidence for the very first time. Despite everyone's amazement (and sometimes their skepticism), the critics of the Copernican model had little argument to offer in response to this newest discovery.

The last of his four central discoveries was to determine that the sun had spots, rather than being some sort of geometrically perfect object the way Aristotle had argued. This one, though, is so much less exciting than the others, but there is one further point worth mentioning. Because of the movements of the dark patches that appeared and then disappeared, Galileo argued that the sun must be rotating on its axis just like other objects.

In contrast, his discoveries had opened up the possibility that, as Bruno (and others) had imagined, there might be other worlds like ours—planets around other stars and moons around those planets. The discovery that the universe was much larger than we thought and that there might be stars we still couldn't see gave people the idea that the universe might really be infinite in scale.

In that larger context, then, we can appreciate that the confirmation and strengthening of Copernicus's model made it more dangerous to the Catholic Church, and Galileo's claims were much harder to overlook than Copernicus's had been. Feeling like it was under a kind of attack on its authority, the church went after Galileo and tried to silence him. In the face of these steep consequences, Galileo had pushed astronomy forward so far and so decisively that others were inspired to take up the cause.

KEPLER MAKES NEW STRIDES

Even while Galileo was in Italy solving puzzles in the night sky with his telescope, Kepler was in Germany working through some of the puzzles in his mind. Kepler realized there was a significant problem with the model for the orbits of the planets. The fact that the new model was actually worse in some ways at predicting the position of planets meant that the competition between the complicated model inherited from Ptolemy and the new, simpler model from Copernicus was far from over. Kepler was the one to solve this piece of the puzzle, and it was a big piece that helped deliver a crucial victory in that competition.

Out of all of them, Mars was the planet that had the biggest gap between observations and what the model predicted. To tackle this, Kepler was able to use the notes gathered by Tycho Brahe, the Danish astronomer who had collected impressively thorough observations. When Brahe died, Kepler was allowed to use this precise data on the positions of the planets.

Kepler worked tirelessly for over ten years trying to fit the observed positions onto one curve after another. His process was made even harder because even just to fit the planets on curves involved making some assumptions about the orbit of Earth itself. Kepler made these assumptions in order to know what to add to or subtract from the other orbits to get the measurements between them.

His work came close any number of times, and many of the efforts to find a circle that would work for Earth and would fit with one that worked for Mars were good enough that earlier astronomers would have been happy with the results. Eventually, he had to give up on finding any combination of circles that could possibly work. At that point, he tried switching to oval shapes instead of circles. Even then, however, he could not close the gaps between his models and Brahe's careful observations.

It was only when he happened, with some luck, to notice that the gaps showed a kind of mathematical pattern that he figured out that the remaining gaps were caused by the fact that the planets did not travel at a steady speed. After a ton of attempts, he finally managed to come up with a solution. The path was an ellipse that had the sun as one of the two focus points. Once he puzzled out this one path, he was able to figure out the different ellipses for the other planets as well.

Unlocking this part of the puzzle also cleared the way for a few other major developments. Kepler was able to formulate, for instance, the other two of his three laws of planetary motion. The shape and focus points were one of these laws. The second was formulated to explain the variation in speed of each of the planets. This law would actually also help to solve one of the small questions we introduced earlier about planets sometimes seeming to go faster than others.

Kepler was able to show that while the speed changed at different points around the orbital ellipse, a string tied from the sun to the planet would sweep out the same area in a given amount of time. Think of it this way: if you had an oval-shaped pie and you gave everyone a piece that weighed the same, some would be long and skinny while some would be short and fat. Kepler was showing that traveling along the edge crust for the skinny pieces, which would be a short distance, takes the same amount of time as traveling along the edge crust for the fat pieces, which would be a longer distance.

His third law didn't seem to have a particularly big impact on understanding the planets and their motions as the first two did, but it is a beautiful piece of reasoning. It also gave a certain kind of joy to Kepler's Neoplatonic heart. The idea is that we have two planets, planet 1 and planet 2. First, let T_1 stand for the length of time it takes planet 1 to orbit the sun, and T_2 is the same for planet 2. Then we let D_1 stand for the average distance from planet 1 to the sun, and D_2 equals the average distance from planet 2 to the sun. If we take T_1 over T_2 and square it, that equals the result when we take D_1 over D_2 and cube it (take it to the power of 3).

It is a sad truth that Kepler's work was dense and could really only be understood by highly trained astronomers. Many of them found his new method to be thinking too far out of the box, so they continued to resist and oppose the new developments. His planetary laws, for example, were not widely accepted until the end of the seventeenth century.

Kepler had, then, pushed us a huge step forward. Neither he nor Galileo could find a way to explain how it could be possible for Earth and the other planets, suspended in space, to travel around and around the sun (either in circles like Galileo

A painting from the early 1700s portrays an older Isaac Newton.

wanted or in ellipses like Kepler wanted) without something pushing them. Kepler's approach was to try to explain it by using magnetism. While it was wrong, it was actually not as bad a theory as it sounds at first. In 1600, a scientist by the name of William Gilbert had proven that Earth was a kind of giant magnet. Kepler generalized the idea by figuring that if the sun also exerted magnetic fields, this could explain the whole thing. It turns out that the sun does exert magnetic fields, even though this is not actually the explanation. For that, we would have to wait for Sir Isaac Newton.

NEWTON and GRAVITY

As much as it might seem that the scientists of the revolution were fighting against religious faith, it is not so simple. Rather, they were struggling to pursue their knowledge even against certain narrow or rigid interpretations of faith. In Newton, we find one of the great intellects and scientists of his age (or of any age, really). For Newton, his work to understand the universe was, like the others, a way to learn something about the god who had created it.

Newton had a genius that was off the charts, so to speak. He demonstrated great insight both as an experimental scientist and as a genius of mathematics. It was because of this particular combination that he was able to carry the Copernican model to the next level. He developed new ideas and experiments in mechanics (what we usually think of as physics) and in the unfolding of the calculus.

At the heart of his method was the strategy of investigating the forces of nature through the motions of things. Then he would develop mathematical models that would allow him to carry the investigation further. His greatest work, published in 1687, had the long Latin title of *Philosophiae Naturalis Principia Mathematica.* The English version would be translated as *Mathematical Principles of Natural Philosophy.* For short, it is usually just referred to as the *Principia*. In it, he laid out a new version of physics that brought both the larger motions of the planets and the smaller worldly objects like cannonballs under one framework. By bringing all of the different recent developments under one shared set of principles, the best ideas of Copernicus, Kepler, and Galileo came into agreement.

At the heart of his new physics were the three laws of motion. Second only to those was his discovery of the principles of gravity. With these elements, Newton thought, God had ordered the movements of the entire cosmos. Throughout his writing, he seems to hold that gravity is a kind of direct act of God, and if not for gravity, his models would predict the collapse of the solar system.

In particular, Newton's law of universal gravitation, as he called it, brought into focus some of the work by Galileo and others on the workings of Earth's gravity. According to his law, the force of gravity between two bodies is proportional to the multiplication of their masses. At the same time, the force of gravity is inversely proportional to the square of the distance between those two bodies. That last part might be confusing, but a quick example can illustrate. If you move two bodies $2x$ as far apart, the gravitational attraction between them is the **inverse** (1/2) squared, so $(1/2)^2$, or 1/4 what it was before.

Through the use of insightful mathematical argumentation, Newton was able to show that the same laws he demonstrated in the laboratory could be used to prove Kepler's three laws of planetary motion.

Another key component of Newton's system was to posit the existence of absolute space and absolute time. In these cases the term "absolute" just means they are sort of apart from anything else. You can think of it as a sort of fixed, three-dimensional grid. Both are static, and everything else moves in relation to this fixed framework. These provide the sort of stage, if you will, for the motions and actions of various bodies. This approach would stand unchallenged until the time of Einstein, and even now it is a more useful way of thinking about things when we are not working with objects traveling the speed of light or being pulled into a **black hole**.

The REVOLUTION BECOMES CONTAGIOUS

The transformed model of how to conduct science would ultimately catch on and spread to other fields of research. Within science, the most immediately relevant application was the growing discipline of **microscopy**. The same lens technology that was used in telescopes was also turned around to be used in microscopes.

Surprisingly, the Industrial Revolution and the new approaches to politics that would develop into the eighteenth and nineteenth centuries also had roots in the new scientific method. The work of Copernicus, and the others, had given us a kind of clockwork universe that runs on mathematical principles. This kind of mentality was to have an influence on great thinkers like the Founding Fathers of the United States. The formative documents like the Constitution and the Declaration of Independence aim at something like a clear, smoothly running clockwork society. They also, however, take a recognizable approach to achieving that goal by building from the ground up as Newton and the others did. Instead of the bodies and laws of motion, we find individuals and rights.

To understand how the Copernican Revolution caught on, we must reflect back on the overall shape of the transformation. The great thinkers of the revolution were driven in a whole new way to follow their deepest intuition. This intuition was that

Antonie van Leeuwenhoek would revolutionize science through the microscope rather than the telescope.

explanations of the world around us must follow the patterns of geometry and mathematics. That is, it is mathematics that leads us in our exploration.

Indeed, Copernicus decided to change the cosmological model first so it would fit with a more elegant piece of geometry. It was only *after* doing so that he was able to begin exploring and testing out just how much better his model would help us to account for observations and to explain away things that seemed like exceptions to the pattern (or irregularities, at least). We also see the importance of mathematics when Le Verrier puzzled out Neptune's existence, and only after that—and because of the mathematics—were Neptune's observers able to confirm his findings.

We will see this same pattern reflected in the fields that were affected by the new method of scientific inquiry. Politics, for instance, would blossom into economics with the addition of mathematic inquiry. Problems and discrepancies within math are often what drive research and discovery.

The Evolution of the Telescope

The invention of the telescope is credited to Hans Lippershey, a Dutch maker of eyeglasses. As the story goes, an apprentice to Lippershey stumbled on the effect of looking through two convex lenses, one in front of the other. He discovered that the magnification compounded. Lippershey saw the potential impact of this finding and spent the next few years working out the proper placement within a wooden tube to create a stable, focused result. In 1608, he sold his design to the Dutch military.

As Galileo and others worked to improve on the design of these refracting telescopes (telescopes that have one lens in front of the other, which the viewer looks straight through), the solution of longer and longer tubes proved problematic. Around 1663, James Gregory, a mathematician from Scotland, had the idea of using a mirror to focus the image for the user. His solution allowed for a shorter telescope (called a reflecting telescope) with a larger circumference. Though he published his design and reasoning, it would fall to Newton to actually construct the first of Gregory's telescopes.

Newton's design built on Gregory's idea and used a primary and a secondary mirror to bring the light into focus. This type of telescope is easily recognized by the viewing lens located on the side of the tube. Since optics was an area of great interest for Newton, it should be little surprise that he was almost as fascinated by the mathematics behind this new design as he was about looking through his telescope at the stars!

This was the first reflecting telescope, and it was built by Newton in 1668.

Our modern portrayal of the solar system is usually simplified somewhat, but it is far beyond the earlier models.

CHAPTER 5

The Influence of the Copernican System Today

As we turn to look at what kind of legacy was created by the Copernican Revolution, we can frame the outcome in terms of a number of small victories and two larger victories. The smaller victories were the particular pieces of the puzzle that our group of geniuses were able to fit into place, such as solving for the changing brightness of the planets and the change in speeds as they orbit the sun.

While those are each valuable steps forward, they are only really a kind of side effect of the two much larger victories that were achieved by the thinkers of the revolution. Those victories have much wider reach in their implications, and they also go down further toward the foundations of science.

FREEDOM from the CHURCH'S CONTROL of SCIENCE

The first of the two major paradigm changes brought about by the Copernican Revolution was to break science out from under the control of the priests and theologians so that scientists could pursue an honest scientific agenda. What that means is they could decide which theories were to be accepted or rejected based on what their observations and calculations determined, rather than having to start with some claims and theories being set in stone and above possible challenge.

Consider, for example, that the only reason the revolution was able to get started at all was because Copernicus was just vague enough in making his claims and his work was just hard enough and boring enough, in a way, to escape wider notice. If his claims had been as strong as Galileo's, by comparison, history would likely have followed a much different path.

It is strange to think of it, but it also probably helped save Copernicus from persecution that he died just as he lit the fuse on the revolution. When he published his book, *On the Revolutions of the Heavenly Bodies* in 1543, he was quite literally on his deathbed. His central claims were radical enough: that the sun is the center, rather than Earth, and that Earth revolves around the sun while spinning on its own axis. However, like all scholars of his era, Copernicus wrote only in Latin, so only a small percentage of people could read it. Perhaps most importantly, his choice of language was very careful. He used phrases like *if* Earth moved, *then* such and such would be the result. Those cautious ways of framing the sentences gave him breathing room against disagreement with religious officials. He could say he did not necessarily believe what the text suggested and that he was only reasoning out how such claims would fit together.

The other thing that helped save him from feeling the immediate wrath of the church is that his research was weaker, so his claims and conclusions felt less dangerous to the investigators from the church. Since there are a number of places where Copernicus's arguments are not well constructed or have flaws, they was not considered a real threat in the way they could have been. Among other things, this meant his book avoided being added to the church's Index of Prohibited Books until seventy-three years after it was first published.

It also happens that one of the church officials who intended to ban the book was prevented by severe illness, and he passed away before he could carry out his plan. On top of that, the **Council of Trent** and the political struggles around it kept the church busy from 1545 to about 1563. By staying out in the open as long as it did, then, Galileo would have the opportunity to read and study it. That, of course, inspired Galileo to take up the investigations himself.

In sharp contrast, Galileo's book was more well constructed and stronger in its claims. The pope had Galileo's *Dialogue on the Two Chief Systems of the World, Ptolemaic and Copernican* brought to a quick halt almost as soon as it was released. Pope Urban VIII pushed the publisher to stop printing any more copies.

The Inquisition demanded that Galileo come to Rome shortly or they would put him in jail and bring him to Rome once his health improved. After eight long months of trial, seven of the ten judges sentenced Galileo to house arrest. The conditions of his house arrest were that he publically announce that his claims were in error, that he would perform religious penance for an extended period of time, and that the Inquisition could still jail him if they decided to.

Eventually, the church showed a little leniency and Galileo was able to live out the last part of his life on his farm. His book stayed on the list of prohibited books long after he had died. In something of a final jab, the pope did not even allow Galileo to be buried on church ground. Instead, Galileo was placed in an unmarked grave in Florence.

It would take ninety-two years before the church permitted his remains to be moved to a better burial place. His book, however, stayed on the index for another ten years, and, in 1744, a version with a lot of editing (they even took out some

Pope Urban VIII was personally responsible for holding back the progress of science for decades.

sections altogether) was released. The more accurate version was not removed from the index until 1835. That was 202 years of suppression … for a science book!

The church would not admit that Earth revolved around the sun until 1820. Then it allowed another astronomer to declare Copernicus's claim about the motion of the Earth as fact. Just think about that for a minute. It took the church two hundred years to admit what Copernicus argued. Understanding this timeline helps give an accurate sense of just what it means to both the practice and the rate of progress that science eventually escaped church control.

ONWARD into the FUTURE

In this final section of the book, then, let us turn to look at how the transformation of the scientific method took root and has continued to bear fruit ever since. It is only fair to mention the prediction of Halley's comet and the discovery of Neptune again. These were both cases where the work the revolutionaries did to give science a strong mathematical foundation shines through.

What follows, then, are some additional examples of the research and discoveries that have parallels to the achievements of the Copernican Revolution. In fact, for some of these cases, it is rather easy to imagine that Newton himself might be running the research and experiments!

The HERSCHELS

Newton died in 1727. Within a few years, the Herschel family (siblings William, Alexander, and Caroline) would take up the life of scientists in the wake of the Copernican Revolution. William was born in Hanover, Germany, in 1738, but moved to England. In midlife he took up astronomy with a passion, spending night after night making careful observations.

William was not content to study closer objects like the sun, the moon, and the planets, though these subjects seemed like enough for most of his contemporaries. Rather, William was captivated by the study of far distant things in the night sky. It is not surprising, then, that one of his major goals was to develop bigger and more powerful telescopes to gain more clarity while looking farther out. In fact, he needed lenses that were larger than lens grinders knew how to make, so he began designing his own. His process also included developing certain kinds of metal, as well.

William's brother and sister both assisted him in his work. This family of astronomers became fairly well known. In 1781, while making a careful survey of the night sky, William happened upon what turned out to be the planet Uranus. This was the first planet discovered since ancient times, and the fame of the Herschels exploded practically overnight.

In addition to discovering two of Uranus's moons, William went on to discover that the blurry **nebulae** were actually composed of individual stars. He is also credited with figuring out a way to determine that the stars themselves actually move. This shattered the ancient belief that the stars were all fixed in their positions. Over the remainder of his career he would discover over 2,000 new nebulae and almost 850 double stars and form a tentative theory about the formation of stars.

William's sister, Caroline, has a remarkable story as well. While she spent most of her time assisting her brother's research and surveys, Caroline also developed into a capable astronomer in her own right. She discovered fourteen new nebulae herself. Notably, she became the first woman to discover a comet, in 1786. Caroline would eventually discover eight comets in all.

The Herschels collaborate on the discovery of Uranus.

Not until the early twentieth century did a scientist (Einstein) push us beyond the ideas of Newton and Copernicus.

EINSTEIN, MATHEMATICS, and the DISCOVERY of BLACK HOLES

Sometimes we do not even receive confirmation of a scientific theory by a later direct observation. At times, proof comes only by means of indirect observations of some kind. In these cases, we have to rely on the math even more than someone like Halley or Copernicus did.

During the development of his general theory of relativity in the early 1900s, Albert Einstein was working on his mathematical equations about the behavior of gravity. The results of this work led him to posit the existence of something like a black hole. The actual term "black hole" was not used until 1967 (by an American physicist named John Wheeler).

Today, the search is ongoing, since we have not yet observed a black hole. Scientists have, however, been able to test out other claims that are predicted by the mathematical model. Therefore, the credibility of the model itself seems to be reliable.

Einstein's Math and General Relativity

In 1915, Einstein's general theory of relativity was published. Then in 1916, physicist and astronomer Karl Schwarzschild of Germany figured out the first set of values that would satisfy Einstein's equations that describe how space-time curves. (Space-time is Einstein's explanation that space and time are not actually separate; they exist together on a continuum.) The solution Schwarzschild came up with gives the shape of the gravitational field that would surround a symmetric, spherical body. We have to remember that a black hole is not a literal hole, as far as we understand it. Rather, it is a star that has collapsed in on itself, creating a density and gravity that are off the charts.

This was not a new idea. As early as the 1790s, an English mathematician named John Mitchell suggested the existence of black holes. Not long afterward, Pierre-Simon Laplace (of France) was able to show that Newton's laws could be used to show the existence of something like an "invisible star." It took some special kinds of calculation to figure out what the mass and size would have to be for a black hole to have such a strong gravitational pull. In fact, a black hole's pull is so strong that something would have to be traveling faster than the speed of light in order to escape it.

Einstein's model, as part of his theory of general relativity, was a meaningful step forward. Until there was a way to test out his model, though, it had only a weak supporting argument. The conditions that would allow for a test did not come around until 1919.

According to Einstein, when light is traveling from a distant star and passes close to our sun it should be "bent" by the pull of the sun's gravitational force. The effect of that bending action would cause us to perceive the wrong location for the star that is giving off the light. To appreciate the idea here, think of the way light passing through a glass of water causes an illusion that affects the bottom part of your drinking straw. It looks like it is moved over from the top part of your straw sticking out of the water.

The reason Einstein's claim about bending light is so hard to test is that the brightness of the sun would drown out our ability to see light passing near our sun. That means we need to measure the effects during an eclipse. There was a good eclipse for that in 1919, so the English astronomer Arthur Eddington (and others) tested Einstein's prediction. When it turned out the way Einstein's model said it would, Einstein became famous on the scale of a rock star overnight.

EXOPLANETS

The ongoing search for exoplanets is another great example of work that could have almost been taken right out of the notebooks of Galileo or Le Verrier. The term "exoplanet" comes from crunching down the words "extrasolar" and "planet." Therefore, "exoplanet" means planets that exist outside our solar system. As of now, astronomers have confirmed the existence of over three thousand exoplanets, and there are about a thousand more waiting to be confirmed.

Finding exoplanets is particularly interesting because there are several different ways to detect when there are planets orbiting a distant star. Perhaps the simplest to explain and imagine is called transit **photometry**, which means measuring the change in light from the star that's caused by a planet passing in front of it. When a planet passes in front of the star, the amount of light we can detect drops a tiny amount. These kinds of observations can help us figure out how long it takes the planet to orbit its star, as well.

On the other hand, there are two problems with this approach. First, if the path of a planet's orbit does not happen to pass across the sun at our level, we are not able to see it. Second, there is a really high rate of false positives, or when scientists seem to detect such a dip but can't confirm it.

A somewhat more clever method is the radial velocity method. This works by detecting the small pull that a planet exerts on the star it is orbiting. This is obviously easier to catch with our instruments when it is a big planet (since it would pull on the star just a little bit harder). Radial velocity is by far the most successful method of those we have used, and the first exoplanet detected by this method was back in 1995.

Light waves have a different **wavelength** if the source is moving toward you than if it is moving away from you. By using very sensitive spectrometers, we can measure small changes in whether a star is moving toward us or away from us. If the planet circles around the back of the star, compared to us, it would tug the star just a small amount away from us. If the planet circles around in front of the star, it would pull the star just a tiny amount toward us. The bigger the planet is, and the smaller the star is, either or both of these could work to increase the amount of tug created. This is how radial velocity works.

Sometimes both the radial velocity method and the transit photometry method can be used together. In those cases, we learn a lot more about the exoplanet.

The Continual Journey Outward

Modern science continues to help us understand the universe. In science's partnership with math, we are finding more and more stars as we go.

If the men like Galileo and Herschel, who worked to expand the limits of what telescopes could do, were able to see the wonders brought to us by the Hubble Space Telescope, they would be astonished.

Because Earth's atmosphere causes so much distortion of the light from distant stars, the Hubble Space Telescope (named after Edwin Hubble) was launched into orbit in 1990. From its position almost 400 miles (643.7 km) above Earth, the HST stretched the distance we could see out into the universe with ten times better resolution and the ability to detect light fifty times fainter than anything we can see from the ground.

Remember how amazing it was for Galileo and his telescope to discover stars where before people thought there was just empty space? When the Hubble was pointed at a seemingly empty patch of the night sky and left with its shutter open for ten days to let it absorb as much faint light as it could, an image called the Hubble Deep Field came back. That was

The Hubble Ultra Deep Field image expanded our conception of the size of the universe.

in 1995, and that one tiny empty patch of sky contained three thousand galaxies that no human had ever seen before. The tiny patch of sky the HST was pointed at was about one-twenty-four-millionth of the sky (yes, a tiny, tiny fraction!). As it turns out, the Hubble was just warming up.

In 2004, a similar study was conducted. This time, the Hubble absorbed light over the course of several months. If you stretched your arm out in front of you holding a piece of paper that was 0.04 inches (1 millimeter) by 0.04 inches, that would be about how big the area was that it focused on this time. The image we got back is called the Hubble Ultra Deep Field, and it contains ten thousand different galaxies. That led to an estimate that the known universe contains approximately two hundred billion galaxies. (Keep in mind that even a midsized galaxy like our own Milky Way contains somewhere between one hundred billion and four hundred billion stars). A 2016 study calculated that it is actually closer to three trillion galaxies.

That seems amazing, and it is hard to even imagine what we might find when the HST is phased out and replaced by the Webb Space Telescope, with a mirror seven times larger than that of the Hubble.

Searching for the God Particle

Talking about stars and planets might make it seem that the Copernican Revolution has carried us outward into space. However, the same transformation of optics and inquiry also led to the birth of scientific research using a microscope, taking us down to smaller and smaller levels of our world.

Present inquiries continue in that direction as well, and they have gone far past the little organisms that scientists like Antonie van Leeuwenhoek first saw through elementary microscopes. One amazing line of research is the quest to find something called the Higgs boson. The Higgs boson is also known as the

God particle because it is what gives something mass. Our mathematical models tell us it must be there—it must exist, but we have not yet found a way to observe it. In fact, the best we have done so far is to try to catch glimpses of its effect on other things. To even accomplish that much, we have had to invent whole new kinds of science equipment that is built to run incredibly complicated kinds of math to detect and evaluate the reactions it just recorded.

The Higgs boson is a really complicated idea to try to explain using analogies, so we can only get parts of the idea across. Here is roughly what is going on with the God particle: imagine you drop objects in a swimming pool. Some objects zip through the water without much resistance—like ball bearings, for instance. Yet other objects get slowed down because of their size—a person, for example. In this scenario, the water is a Higgs field, and each little water molecule is a boson. However, we must imagine that water molecules are constantly popping in and out of existence. Now the resistance—how hard it is to move through the water—is determined not actually by our size, but by our mass. In short, this is what today's physicists are working on.

And the Higgs boson is such a great example of the tree Copernicus planted bearing fruit. We can use math and new tools built with that math to observe light that has been traveling from its home galaxy for billions of years. We can also leverage the mathematical inheritance we have received from the Copernican Revolution to try to observe things like this particle that decay almost instantly after being created.

The Search for Planet 9

In recent studies of Earth's solar system, astronomers have found a number of irregularities that suggest there is another planet in the outer range of our solar system. Remember how Le Verrier found Neptune because Uranus was acting strange?

The exact same idea is going on here, except the things that are acting strange are dwarf planets.

We are talking about objects that are way out in the outer part of our solar system, past Neptune and the Kuiper Belt. Using the gifts Newton and the others gave us, the researchers are able to posit certain characteristics of the planet in order to account for the strange behaviors. The prediction that comes back according to math is a planet that would have about ten times the mass of Earth and a diameter between two and four times the size of Earth.

String Theory

String theory really makes the best example to save for last because here we are well beyond anything even sort of observable. In fact, that is one of the criticisms of string theory in physics—we don't have anything but mathematical models to challenge the core ideas of the model.

In some ways, this mystery is similar to that of trying to find the Higgs boson. In both cases, science is trying to reach down to the smallest levels of the material universe and understand what everything is made of. The key difference is that for string theory, it is mathematical all the way down. To see what that means, we need to get a sense of what string theory claims.

The central idea is that instead of some kind of teeny, tiny particle (like the way electrons and neutrons are pictured as balls) being the smallest building blocks of everything, the smallest level is more like a folded string that is vibrating. This would be sort of like the way that strings of musical instruments vibrate. The particular rate of vibration of a given string would determine what mass and charge that string would have. So, for example, one rate of vibration would be an electron (what atoms are made of). If the string was vibrating at a different frequency, it would be a quark (what protons and neutrons are made of), and so on.

As we begin to dig down a little farther, we realize that calling these things strings is an oversimplification. The strings in string theory have many more dimensions than we are able to think of without using some seriously advanced mathematics. We can get a little bit of an idea of what is going on here, though. We can think of one-dimensional objects, of course, and you probably remember those from geometry. The model is a line. To be truly one-dimensional, a line would have to have a thickness of zero, of course. To represent two dimensions, then, we can imagine a grid with one set of lines going horizontally and one vertically. Picture a piece of graph paper.

Now, if we lay the graph paper down, we have the same kind of picture as if we were to look at the grid of streets in a city. To get three dimensions, we would add height. A good way to picture this would be to think of going up in an elevator in different buildings. Everything is good up to this point, and we think of 3D objects in our everyday lives all the time. To think of what a fourth dimension might be, just consider that after we go up to floor 21 (or whatever), we then travel some distance along the hallway. How far we go would represent the different possible values for dimension four. Are you still with me?

After this, it starts getting messy, and trying to make a real-world picture gets about as much stuff wrong as it gets right. We get a little bit further in understanding the model, though, if you allow this little imaginary section of a city to bend around and connect the ends together to make a loop.
We achieved four dimensions, but, here's the mind-blowing part: the mathematical equations actually show that these strings would contain nine dimensions (or more)! Not to mention the unbelievably mathematical idea that everything might be made out of music, metaphorically speaking.

This is the reflection nebula Messier 78.

BACK to BRUNO

That has been quite a tour through history and the solar system since we talked about Giordano Bruno in the introduction. When we started out with Bruno and his dream vision of an infinite universe, we suggested that the scientists we would meet were trying to stretch themselves and the tools for expressing ideas about the universe.

We can certainly see the ways they did that. Again and again, they solved pieces of the puzzle so they could reach to fit more into their model: new telescopes, new theories of gravity, new kinds of math, and so on. Now we must stop and ask, as we would of any great artist or musician: Did they not only stretch themselves but also carry us with them? Has their accomplishment helped *us* see a larger universe by looking at it through different lenses?

Imagine what new dream might have been sparked for Bruno if he could have looked at the universe through the lenses of the Hubble Telescope.

Henrietta Swan Leavitt

It is easy to form a skewed perspective of scientific brilliance when reading about breakthroughs like the Copernican Revolution. As we saw with Caroline Herschel, some of the most vital steps along the way were made by women.

Henrietta Swan Leavitt (1868–1921) is a wonderful example of this. Leavitt, an American scientist, discovered four novas and more than four hundred variable stars (which means about half of all the ones known at that point) while working on a project at Harvard. Even more amazingly, she figured out that the cycle of changing brightness in a special group of stars matched up to their actual luminosity. Luminosity means how much light something gives off in a given period—what we *normally* think of as brightness.

In particular, she was able to show that the ones whose (seeming) brightness goes up and down at a slower rate are the ones actually giving off more light. This was important because before her discovery, there was no dependable way to tell if star A seemed brighter than star B because it gave off more light or because it was just closer to us.

By matching up the length of the cycle and the luminosity, Leavitt's work allowed astronomers to gauge, for the first time, the size of the Milky Way galaxy. Not only that, she made it possible for Hubble to figure out the distance to the Great Nebula in the Andromeda galaxy. That was the first distance measured to something outside the Milky Way.

Henrietta Swan Leavitt's astronomical mind was as brilliant as the stars she studied.

Chronology

310 BCE–230 BCE The lifespan of Aristarchus of Samos

100 CE–170 CE The lifespan of Ptolemy

1435 Leon Battista Alberti publishes *On Painting*

1473–1543 The lifespan of Nicolaus Copernicus

1514–1574 The lifespan of Georg Rheticus

1539 Copernicus begins to work with Rheticus

1548–1600 The lifespan of Giordano Bruno

1564–1642 The lifespan of Galileo Galilei

1571–1630 The lifespan of Johannes Kepler

1584 Bruno publishes *On the Infinity of the Universe*

1596 Kepler publishes *The Cosmographic Mystery*

1609 Galileo discovers mountains and craters on the moon

1610 Galileo discovers the four largest moons of Jupiter

1642–1727 The lifespan of Isaac Newton

1656–1742 The lifespan of Edmond Halley

1687 Newton publishes *The Mathematical Principles of Natural Philosophy*

1781 William Herschel discovers the planet Uranus

1786 Caroline Herschel becomes the first woman to discover a comet

1799–1825 Pierre-Simon Laplace publishes *Celestial Mechanics*

1820 The Catholic Church admits that Earth revolves around the sun

1835 The Catholic Church lifts its ban on Galileo's book

1846 Urbain Le Verrier discovers the planet Neptune

1919 Albert Einstein's theory of relativity is proven during an eclipse

1923 Edwin Hubble discovers that there are galaxies beyond our own

1925 Cecilia Payne-Gaposchkin proves stars are like the sun

1990 The Hubble Space Telescope is launched into low Earth orbit

1995 First exoplanet is discovered around a sun-like star: Pegasi 51b

2013 Preliminary confirmation of detection of Higgs boson

2014 A previously undiscovered, large ninth planet in our solar system is proposed

Glossary

astronomer A person who studies objects in space, including planets, stars, moons, galaxies, nebulae, and so on.

black hole A collapsed star with such a high density that its gravitational field is too strong for objects or radiation (including light) to escape its pull.

calculus A special branch of mathematics that deals with the behavior of change—such as the motion of objects—and does so through calculations and properties of infinite series and infinitesimal amounts.

celestial Having to do with the sky beyond Earth's atmosphere, space, and objects in space.

comets Celestial objects that orbit the sun on elongated orbital paths and contain a core of ice and dust and give off a gaseous trail as they travel.

Copernican Revolution The widespread adoption of Copernicus's idea that the sun, not Earth, is at the center of the solar system.

cosmologist Someone who studies the origin and development of the universe.

Council of Trent A meeting of all the major leaders of the Catholic Church to make decisions about policies and official Catholic teachings as well as to decide which claims by Protestants went against Catholic teachings. It lasted from 1545 to 1563.

deferent In Ptolemaic astronomy, this is the circle around Earth along which a given celestial body (like a planet) moves.

epicycle A small circle that moves along the edge of a larger circle.

friction The force that interferes with motion when two surfaces rub or slide against each other.

geocentrism An understanding of the universe that places Earth at the center.

gravity The force pulling objects with mass toward each other.

heliocentrism An understanding of the universe that places the sun at the center.

inclined planes Surfaces with a slope that is not zero degrees.

infinite Having no limit in one or more ways, such as extending without end in time or in space; also said of a series of countable objects having no final number.

inverse The opposite; in math the multiplicative inverse expresses a number as a fraction and then flips it over so that multiplying the two would give an answer of x/x, which equals 1.

mass The property of matter that makes it resist being moved; the amount of mass is how much of something there is—can be approximately thought of as an object's weight when being affected by gravity.

microscopy The field of studying things with a microscope; used to cover a wide range of scientific areas of investigation.

nebulae Celestial clouds of dust and gases; some nebulae have ingredients and conditions that result in the formation of new stars.

optics The branch of science that deals with the behavior of light; includes study of lenses, mirrors, sight, telescopes, and so on.

paradigm The clearest or most representative example of something; can also mean the collection of beliefs and ways of looking at or thinking about some topic or area of study.

photometry The science that deals with the measurement of light, including wavelengths and brightness.

physics The branch of science dealing with matter and energy; including the interactions that cause force and motion.

Protestant Reformation This was the period from 1517 to 1648 when several groups broke away from the Catholic Church over various differences in belief about religion and about the role and function of a church. The leaders of the split include Martin Luther, John Calvin, and Huldrych Zwingli.

refracting A descriptor of bending of light waves as they pass from one medium to another, as from air into a lens, or from the lens back out.

relativity Several closely related scientific discoveries showing that certain things behave differently depending on where the observer is in relation to the other objects or forces.

retrograde Moving backward in direction or in time.

slope A surface where one side is higher than the other; also used to refer to a measurement of how steep such a difference is.

sunspots Dark areas on the surface of the sun that occur when the magnetic fields cause temporary areas of cooler temperatures.

tangent A line that rests against the edge of a circle or curve so they would touch at only one possible point; also refers to a central calculation in trigonometry.

trigonometry A branch of mathematics focusing on the properties of triangles, circles, and cyclical patterns or motions—like tides, population cycles, and so on.

velocity How fast something is traveling and in which direction.

wavelength The distance between high points of an electromagnetic wave, like light or sound.

Further Information

BOOKS

Danielson, Dennis. *The First Copernican: Georg Joachim Rheticus and the Rise of the Copernican Revolution.* New York, NY: Walker & Co., 2006.

Farndon, John. *The Great Scientists: From Euclid to Stephen Hawking*. London, UK: Arcturus, 2005.

Pruett, David. *Reason and Wonder: A Copernican Revolution in Science and Spirit*. Westport, CT: Praeger, 2012.

Repcheck, Jack. *Copernicus' Secret: How the Scientific Revolution Began.* New York, NY: Simon & Schuster, 2007.

WEBSITES

Neo K12

http://www.neok12.com/Universe.htm

This website has a variety of games, videos, and lessons on different astronomy topics.

Nova: Inventing Telescopes
http://www.pbs.org/wgbh/nova/tech/inventing-telescopes.html

PBS's website has interesting, brief articles on Galileo's experiments—some of which are interactive.

VIDEOS

"From Hidden to Modern Figures"
https://www.nasa.gov/modernfigures/videos

NASA's video series features career advice from women working in aerospace, including astrophysicists and engineers.

"Galileo's Battle for the Heavens"
http://www.pbs.org/wgbh/nova/ancient/galileo-battle-for-the-heavens.html

This two-hour video from NOVA explores Galileo's struggles against the Catholic Church.

"Isaac Newton"
http://www.watchknowlearn.org/Category.aspx?CategoryID=7953

The St. Charles Place Education Foundation hosts a collection of short videos on Newton and his major ideas and discoveries. The menu on the left also links to collected videos on other thinkers, like Galileo and Einstein.

"Isaac Newton and a Scientific Revolution"
http://www.history.com/topics/enlightenment/videos/isaac-newton-and-a-scientific-revolution

Explore the link between Newton's work and the Protestant Reformation.

"The Sound of Black Holes Colliding Proves Einstein Was Right"
http://www.pbs.org/video/2365666125/

Watch an easy-to-follow explanation about the first evidence of gravitational waves.

Bibliography

Alexander, Amir. *Infinitesimal: How a Dangerous Mathematical Theory Shaped the Modern World.* New York, NY: Farrar, Strauss & Giroux, 2015.

Bardi, Jason Socrates. *The Calculus Wars: Newton, Leibniz, and the Greatest Mathematical Clash of All Time*. New York, NY: Basic Books, 2007.

Bauer, Susan Wise. *The Story of Science: From the Writings of Aristotle to the Big Bang*. New York, NY: W.W. Norton and Company, 2015.

Bienkowska, Barbara, ed. *The Scientific World of Copernicus.* Boston, MA: Reidel Publishing, 1973.

Burke, James. *The Day the Universe Changed: How Galileo's Telescope Changed the Truth.* New York, NY: Back Bay Books, 1995.

Dobrzycki, Jerzy, ed. *The Reception of Copernicus' Heliocentric Theory*. Boston, MA: Reidel Publishing, 1973.

Dolnick, Edward. *The Clockwork Universe: Isaac Newton, the Royal Society, and the Birth of the Modern World*. New York, NY: Harper Perennial, 2012.

Gingerich, Owen. *Copernicus: A Very Short Introduction*. New York, NY: Oxford University Press, 2016.

Goble, Todd. *Nicholas Copernicus: And the Founding of Modern Astronomy*. Greensboro, NC: Morgan Reynolds Publishing, 2004.

Gribbin, John. *The Scientists: A History of Science Told Through the Lives of Its Greatest Inventors*. New York, NY: Random House, 2004.

Iliffe, Rob. *Newton: A Very Short Introduction*. New York, NY: Oxford University Press, 2007.

Johnson, George. *Miss Leavitt's Stars: The Untold Story of the Woman Who Discovered How to Measure the Universe*. New York, NY: W.W. Norton & Company, 2006.

Koyré, Alexandre, and R. E. W. Maddison, translator. *The Astronomical Revolution*. Ithaca, NY: Cornell University Press, 1973.

Kuhn, Thomas. *The Copernican Revolution*. Boston, MA: Harvard University Press, 1957.

———. *The Structure of Scientific Revolutions*. 3rd edition. Chicago, IL: University of Chicago Press, 1996.

Lindberg, David C. *The Beginnings of Western Science: The European Scientific Tradition in Philosophical, Religious, and Institutional Context, Prehistory to A.D. 1450*. 2nd edition. Chicago, IL: University of Chicago Press, 2008.

Long, Pamela O. *Artisan/Practitioners and the Rise of the New Sciences, 1400–1600*. Corvallis, OR: OSU Press Horning Visiting Scholars Publication, 2011.

Love, David. *Kepler and the Universe: How One Man Revolutionized Astronomy*. Amherst, NY: Prometheus Books, 2015.

Miller, Ron. *Recentering the Universe: The Radical Theories of Copernicus, Kepler, Galileo, and Newton*. Minneapolis, MN: Twenty-First Century Books, 2013.

Sobel, Dava. *A More Perfect Heaven: How Copernicus Revolutionized the Cosmos*. New York, NY: Bloomsbury, 2012.

———. *The Glass Universe: How the Ladies of the Harvard Observatory Took the Measure of the Stars*. New York, NY: Viking, 2016.

Stokes, Mitch. *Isaac Newton*. Nashville, TN: Thomas Nelson Publishing, 2010.

———. *Galileo*. Nashville, TN: Thomas Nelson Publishing, 2010.

Thorne, Kip. *Black Holes and Time Warps: Einstein's Outrageous Legacy*. New York, NY: W.W. Norton & Company, 1995.

Wootton, David. *The Invention of Science: A New History of the Scientific Revolution*. New York, NY: Harper Collins, 2015.

———. *Galileo: Watcher of the Skies*. New Haven, CT: Yale University Press, 2013.

Index

Page numbers in **boldface** are illustrations. Entries in **boldface** are glossary terms.

About the Author

Erik Richardson is an award-winning teacher from Milwaukee. While he is on sabbatical from teaching middle school math and science, he is teaching a variety of college courses. In addition to writing in his free time, he also runs a small nonprofit called Every Einstein that works to provide STEM resources for students and teachers around the country.